# A PRACTICAL GUIDE
# TO THE STUDY OF
# THE PRODUCTIVITY OF
# LARGE HERBIVORES

IBP HANDBOOK No. 7

# A Practical Guide to the Study of the Productivity of Large Herbivores

Edited by

FRANK B. GOLLEY

and

HELMUT K. BUECHNER

INTERNATIONAL BIOLOGICAL PROGRAMME

7 MARYLEBONE ROAD, LONDON NW1

BLACKWELL SCIENTIFIC PUBLICATIONS

OXFORD AND EDINBURGH

SBN 632 05530 8

FIRST EDITION 1968

*The programme of work leading
to the publication of this
Handbook has been supported
in part by grants to the
International Council of
Scientific Unions made by the
Ford and Nuffield Foundations*

*Printed and bound in Great Britain by*
BURGESS AND SON (ABINGDON) LIMITED
ABINGDON, BERKS

# Contents

# Contributors

**A. T. Bergerud** University of Victoria, Victoria, British Columbia, Canada

**R. C. Bigalke** Natal Parks, Game and Fish Preservation Board, Pietermaritzburg, South Africa

**Harold H. Biswell** School of Forestry, University of California, Berkeley, California, U.S.A.

**I. Lehr Brisbin** Institute of Ecology, University of Georgia, Athens, Ga., U.S.A.

**Helmut K. Buechner** Office of Ecology, Smithsonian Institution, Washington, D.C., U.S.A.

**C. S. Christian** C.S.I.R.O., Canberra City, Australia

**Richard G. Clements** Institute of Ecology, University of Georgia, Athens, Ga, U.S.A.

**I. McT. Cowan** University of British Columbia, Vancouver, B.C., Canada

**W. J. Eggeling** Nature Conservancy, Edinburgh, U.K.

**Frank E. Egler** Aton Forest, Norfolk, Conn., U.S.A.

**Frank B. Golley**  Institute of Ecology, University of Georgia, Athens, Ga., U.S.A.

**A. M. Harthoorn**  University College, Nairobi, Kenya

**Harold F. Heady**  University of California, Berkeley, Calif., U.S.A.

**P. N. Hobson**  Rowett Research Institute, Bucksburn, Aberdeen, U.K.

**R. R. Hofmann**  University College, Nairobi, Kenya

**C. B. Huffaker**  Division of Biological Control, Department of Entomology, University of California, Albany, Calif., U.S.A.

**B. C. Jansen**  Veterinary Research Institute, P.O. Onderstepoort, South Africa

**R. N. B. Kay**  Rowett Research Institute, Aberdeen, U.K.

**Allen Keast**  Biology Department, Queen's University, Kingston, Ontario, Canada

**V. A. Kovda**  Department of Pedology, Moscow State University, U.S.S.R.

**W. V. Macfarlane**  Waite Agricultural Research Institute, Adelaide, South Australia

**S. O. Mann**  Rowett Research Institute, Bucksburn, Aberdeen, U.K.

**G. R. Moule**  Australian Wool Board, Sydney, Australia

**P. N. O'Donoghue**  Royal Postgraduate Medical School, London, U.K.

**Kenneth W. Parker**  Forest Service, U.S. Department of Agriculture, Washington, D.C., U.S.A.

**George A. Petrides**  Departments of Fisheries and Wildlife, and Zoology, Michigan State University, East Lansing, Michigan, U.S.A.

**A. T. Phillipson**  Department of Veterinary Clinical Studies, School of Veterinary Medicine, Cambridge, U.K.

**U. de V. Pienaar**  Kruger National Park, South Africa

**N. W. Pirie, F.R.S.**  Rothamsted Experimental Station, Harpenden, U.K.

**R. L. Reid**  Hill Farming Research Organisation, Edinburgh, U.K.

**Henry L. Short**  Southern Forest Experimental Station, Forest Service, U.S.D.A., Nacogdoches, Texas, U.S.A.

**R. Summers**  Rowett Research Institute, Bucksburn, Aberdeen, U.K.

**J. H. Topps**  School of Agriculture, University of Aberdeen, U.K.

**N. Newton Turner**  C.S.I.R.O., Byde, New South Wales, Australia

**George M. Van Dyne**  College of Forestry and Natural Resources, Colorado State University, Ft. Collins, Colorado, U.S.A.

**Robert E. Williams**  Soil Conservation Service, Washington, D.C., U.S.A.

**A. J. Wood**  University of Victoria, Victoria, B.C., Canada

**I. V. Yakushevskaya**  Department of Pedology, Moscow State University, U.S.S.R.

**N. T. M. Yeates**  University of New England, Armidale, New South Wales, Australia

# Section I

# 1
# Introduction

FRANK B GOLLEY and HELMUT K BUECHNER

At Aberdeen and Cambridge, United Kingdom, a meeting was held in September 1965 to discuss the study of the productivity of large herbivores under auspices of the International Biological Program (IBP). One of the objectives of this stimulating meeting was the preparation of the synopsis for a practical guide to the study of large herbivores. Large herbivores, within the context of this IBP program, include both domesticated and undomesticated terrestrial mammals; and, therefore, the agricultural, range management, wildlife management, physiological, veterinary and ecological disciplines were represented at the meeting. Seldom do individuals representing such a range of scientific interests focus their attention on a single problem and it is this combination of diverse viewpoints that is both the problem and opportunity for IBP. We all work in the belief that the new information gathered through IBP and the increased contacts between scientists will help lead toward ecological stability of the world ecosystem.

We hinted at the problems of organizing and carrying out an IBP program. Nowhere have these been more apparent than in the preparation of this guide. The project was inordinately delayed through unavoidable conflicts with other duties of the original editor (Buechner) and many of the authors of the sections. In February, 1967, Golley accepted co-editorship of the volume and the present organization is his responsibility. The original synopsis consisted of the following chapters:

1. Basic surveys
2. Primary and secondary productivity
3. Numbers and diversity
4. Categories and age determination
5. Reproduction
6. Post natal growth
7. Life tables
8. Environmental physiology

9. Plant utilization
10. Nutrition
11. Predators
12. Epizootiology
13. Animal productivity
14. Use, management and husbandry

This schema reflects an orientation to animal husbandry and the applied animal sciences. Insufficient manuscripts were available in February to fulfill the outline of the synopsis and, therefore, a new synopsis was prepared based largely on those manuscripts at hand. The new schema made energy flow and secondary productivity the central theme followed by management considerations, with management conceived in a very broad sense.

Energy flow in animal and plant populations is a central theme of the Terrestrial Productivity section of I B P, and it is appropriate that this theory also provide the framework for the present guide.

No attempts were made to change authors' style or punctuation. The author of each piece is fully identified, and he or she stands responsible for the accuracy of his statements and references. The editors have imposed some uniformity of organization on the papers and have tried to correct the obvious typographic errors.

The guidebook begins with a description of the energy flow concept, followed by more detailed discussion of census techniques and age determination and the components of energy flow; energy ingestion, assimilation, metabolism, and secondary production. Information on a diversity of animals divided into energetically significant categories is preliminary to investigation of their energy dynamics and for this reason is discussed first. The major portion of the guide is combined in a section titled, Management for Secondary Production. This section includes discussion of handling animals, environmental physiology, reproduction, nutrition, habitat manipulation and epizootics. The guide ends with a collection of conversion factors. In many instances methods of study are outlined in detail. In other cases, the article serves as an introduction to the field and gives references to the appropriate literature. We realize that the treatment of the subject is uneven and that our choice of organization may not be useful to all investigators. However, we feel that further delay would radically reduce the usefulness of the guide. We conceive of the guide as a point of departure; let us hope that the omissions or lack of proper emphasis will stimulate research to the extent that a greatly increased understanding of the role and potentialities of large herbivores will emerge as the I B P program matures.

## Acknowledgements

Several persons have been especially helpful in bringing this Guide to completion. We wish to acknowledge the editorial assistance of Elaine Mahoney and Priscilla Golley, and the clerical help of Rosemary Priest and Sharon Crouch. Preparation of the final manuscript was supported by the Institute of Ecology, University of Georgia.

# Section II

# 2

# Energy Flow and Secondary Productivity

GEORGE A PETRIDES, FRANK B GOLLEY
and I LEHR BRISBIN

In the large-herbivore phase of productivity studies, there are three main objectives: (a) to determine the rate, extent and manner of food-energy utilization by the population under study, (b) to estimate the efficiency with which the population converts the energy consumed to the production of animal tissues, and (c) to compare these conversion and efficiency rates with those of other herbivores and/or ecosystems. By definition, herbivores are the first or primary consumers of plant materials. Protoplasmic production in these animals constitutes a portion of secondary production by consumer organisms.

In most cases, secondary-productivity studies will relate to populations at existing densities, eating their normal diets in their usual environments. The extent to which a population has achieved its optimum sustained density (the carrying capacity) in the habitat studied should be noted, however, since this influences productivity. Evidence should be collected leading to a stated conclusion that the herbivores studied occurred as an underpopulated, over-stocked, or optimum population. Most often such evidence would confirm the successional status and relative permanence of the biotic community studied.

Also, it may be important to determine the kinds of food which individual species are capable of assimilating and their needs for particular nutrients. Basic to these and to all productivity (rate of production) investigations, however, are appraisals of energy requirements. In studies of terrestrial herbivores, it is usually desirable to compute production rates in kilocalories per square meter per year.

**The Energy flow concept.** Energy flow is defined as the rate of energy transfer between elements of an ecological system. For consumer organisms, energy flow is equal to the rate at which energy is assimilated from ingested foods. In herbivorous mammals, energy flow may be measured as the difference per unit time between the number of food calories consumed and the amount of calories lost as solid and gaseous wastes or as the calories lost in heat production plus those stored in secondary production.

The total food energy consumed is not all used by an animal Figure 2.1); certain portions are not absorbed, becoming defecated wastes. Of the energy absorbed through the gut wall, some is utilized in the digestion process but most is assimilated into the body tissue. These materials then become

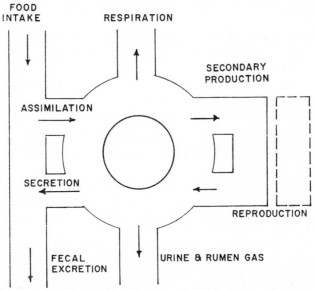

Figure 2.1. Diagram of energy flow through an individual mammal or a population. Redrawn from Davis and Golley, 1963.

available for oxidization through respiration to yield energy for life processes. At each successive energy conversion, that is with absorbtion, assimilation, and oxidation, there is a loss of energy as heat. In addition, some energy absorbed is yielded to the environment as urine and, to a minor degree, as body secretions. Protoplasmic increments which are not required for body maintenance constitute growth or secondary production.

The energy flow concept is a common denominator permitting direct comparisons of standing crops, total production, and rates of production for various species and communities. The principal data required in productivity studies involve the food energy available, consumed, defecated, and utilized in maintenance and body growth. More particularly, it is necessary to know: (1) the food energy available for consumption, (2) the food energy consumed,

(3) the energy absorbed through the gut wall (and mainly assimilated into body tissue), (4) the assimilated energy used in body maintenance, and (5) the energy stored as tissue growth.

1. *Available food energy.* All production studies of herbivores under IBP are planned to be conducted on areas where the primary production is determined also. In practice, this work may be done by zoologists. Methods for determining plant production (mostly by harvest through repeated cuttings) have been published in other IBP handbooks. Since the efficiency of energy conversion by herbivores may require knowledge of the food energy available, it is important that forage production be appraised.

2. *Food energy consumed.* The amounts of food consumed by animals in the wild is difficult to appraise, but possible methods are indicated in the sections following. In an earlier study (Petrides and Swank, 1965), there seemed to be a basis for estimating that half of the average amount of food materials in the rumen equals the food consumed daily by ruminants. In most studies, however, the extent of food consumption must be determined for captive animals, with dry weights and caloric values of foods eaten being determined from random dietary samples.

3. *Assimilation.* Several methods may be used to determine the rate of assimilation. The most common, and generally the most practical, involves careful measurement of the quantities and energy values of foods ingested and of materials egested per time period. The difference between the energy consumed and that defecated is the quantity assimilated. It is impractical, if not impossible, to study the nutrition of complete populations in the wild. Even the calculation of energy requirements for representative individuals normally can be undertaken only for captive animals, and since their energy requirements are less than those of active wild specimens, some correction of data for captives normally is required. Graham (1964), for instance, observed that normally-grazing sheep have energy requirements which are 40% above those of caged sheep. Also, among materials defecated, the determination of discharges of fermentation gases should not be disregarded as it is considerable in at least some ungulates (Benedict, 1936; Brody, 1945).

A method which offers some degree of correction for data derived from captive animals seems to be possible, at least for some populations. Both the amount of food consumed for each unit of feces defecated and the amount of night-time dung defecated as compared with daytime droppings can be ascertained for captive animals. Then, where individual animals are easily observed and their droppings readily located, the collection of daytime feces from wild or semi-wild specimens could provide weight and caloric data which, when multiplied by the previously derived conversion ratios, would provide estimates of total day-night consumption and defecation rates for active animals on wild diets. An alternative procedure might be to fit fecal collection bags to animals which can be recaptured, perhaps semi-wild specimens in large natural paddocks. In many of these studies too, indicators such as chromium-51 can serve as a radioactive marker of foods eaten and residues defecated (Petrides, 1967; Mautz and Petrides, 1967), enabling ready determination of feed: feces ratios.

4. *Energy cost of metabolism.* As yet we cannot measure directly the energy cost of metabolic activities under field conditions. This value usually is derived by subtracting the energy stored as growth from the energy absorbed.

In the absence of nutritional data for the species under investigation, it is possible to extrapolate from a weight-metabolism curve for mammals (*see* Klieber, 1961), since basal metabolism of mammals is related to body size. Even though the calculation is approximate it is useful since, with the data on tissue production, it provides a cross-check on the estimate of energy assimilation.

5. *Energy stored as protoplasmic growth.* Net production by organismic populations is achieved through reproduction and growth. Three sets of information are needed in order to determine the extent of new biomass production on a given site: (a) population density, (b) population age structure, and (c) average individual weight-age relationships.

Methods for determining population density vary from area to area and from species to species. They are reviewed in a later section.

The age characteristics of herbivores also differ by species. Where a cross-section of the population can be viewed, however, even in the absence of established age criteria, recognizable size categories may be designated. These later can be related to age through associated growth studies of captive animals or, preferably, of marked wild specimens.

Where specific size-age criteria are not known, it may be useful to record observed ungulates as calves, immatures, subadults, and adults. The mean age of the relatively newborn calf class usually can be estimated closely.

Often, too, the average age of puberty can be determined from zoo records. Sexual maturity is first reached at subadault size. The immature class represents the ages between that of calves and of subadults. Beyond this, the maximum age of adults usually may be estimated from zoo records (*see* beyond).

An example of this technique was reported by Petrides and Swank (1966) for the African elephant. From the proportion of the population occurring in the several size categories plus other information, it seemed possible to construct a realistic life table for the elephant population. From this table, and with related census and age-weight data, an estimate of the net productivity of thep opulation was determined (Table 2.1). This method also would seem to have applicability for many other species but, in compiling a life table from size-category data, several assumptions are necessary. In general, for large herbivores these are that: (1) the greatest mortality rates are suffered by the very young and the very old, (2) changes in survival rates are gradual, and (3) the maximum age of animals in the wild is not widely divergent from that of zoo specimens though attained by a smaller percentage of the wild population.

The first two conditions seem to prevail normally in large mammal populations. And for maximum longevity, in the African elephant it seemed reasonable to assume that the maximum age of the species in captivity might be equalled by one animal in a thousand in the wild. For other species, too, there are data which indicate that this is not too unlikely a relationship. Pending properly determined data, it is suggested that a rule-of-thumb might be adopted which assumes that ·001 of newborn ungulates reaches the maximum recorded age of captive specimens. Though an unproven ratio, the margin of accuracy for this datum is not required to be close.

Using these assumptions, it is convenient to establish a life table based on an original population of 1000 newborn animals. First, a sequence of annual survival rates is estimated which accords with the stated assumptions and which yields proportions of individuals in each size category that are similar to those observed in the field (Table 2.1).

After a suitable life table (columns 1–5) has been developed, age-weight data must be available which can be used in combination to calculate the population weight gain year by year (columns 6–8). The total weight-gain of the entire population then can be divided by the total number of animals alive to ascertain the net production per animal per year. The ratio between the total population weight-gain and the population biomass per square

*Chapter 2*

TABLE 2.1. Elephant life table and production data.
Queen Elizabeth National Park, Uganda, November, 1956—June, 1957.
From Petrides and Swank, 1966.

| Age$_x$ | Number alive at beginning of Age$_x$ | Survival Rate (s) | Median Number alive | Numbers & proportions in size groups*** | Weight average (pounds) | Weight increment (pounds) | Population Wgt. increment (lbs.) (4) × (7) |
|---|---|---|---|---|---|---|---|
| 1 | 1,000 | ·70 | 850·0 | 1480·0 | 200 | 200 | 170,000 |
| 2 | 700 | ·80 | 630·0 | =calves | 450 | 250 | 157,500 |
|   |     |     |       | = ·143 |     |     |         |
| 3 | 560 | ·90 | 532·0 |        | 700 | 250 | 133,000 |
| 4 | 504 |     | 478·5 |        | 1000 | 300 | 143,550 |
| 5 | 453 |     | 430·5 | 2490·5 | 1350 | 350 | 150,675 |
| 6 | 408 |     | 387·5 | =immatures | 1750 | 400 | 155,000 |
| 7 | 367 |     | 348·5 | = ·237 | 2200 | 450 | 156,825 |
| 8 | 330 |     | 313·5 |        | 2650 | 450 | 141,075 |
| 9 | 297 |     | 282·0 |        | 3100 | 450 | 126,900 |
| 10 | 267 | ·95 | 260·5 |        | 3550 | 450 | 117,225 |
| 11 | 254 |     | 247·5 | 1453·5 | 4000 | 450 | 111,375 |
| 12 | 241 |     | 235·0 | =sub adults | 4450 | 450 | 105,750 |
| 13 | 229 |     | 223·5 |        | 4850 | 400 | 89,400 |
| 14 | 218 |     | 212·5 | = ·138 | 5250 | 400 | 85,000 |
| 15 | 207 | ·98 | 205·0 |        | 5700 | 450 | 92,250 |
| 16 | 203 |     | 201·0 |        | 6150 | 450 | 90,450 |
| 17 | 199 |     | 197·0 |        | 6600 | 450 | 88,650 |
| 18 | 195 |     | 193·0 | 5063·5 | 7000 | 400 | 77,200 |
| 19 | 191 |     | 189·0 | =adults | 7450 | 450 | 85,050 |
| 20 | 187 |     | 185·0 | = ·482 | 7900 | 450 | 83,250 |
| 21 | 183 |     | 181·0 |        | 8250 | 350 | 63,350 |
| 22 | 179 |     | 177·0 |        | 8500 | 250 | 44,250 |
| 23 | 175 |     | 173·0 |        | 8700 | 200 | 34,600 |
| 24 | 171 |     | 169·5 |        | 8900 | 200 | 33,900 |
| 25 | 168 |     | 166·5 |        | 9000 | 100 | 16,650 |
| 26—27 | 3107* | ·98—·80 | 3026·5* |   | 9000 | 0 | 0 |
| Totals | 10,993 |  | 10,487·5 | 1·000 | (5041)** | 9000 | 2,552,875 |

\* Sums of the numbers of animals in each year-class for ages 26–67.
\*\* Average body weight. A constant body weight was assumed for grown animals in the wild though captives have shown an increase.
\*\*\* Proportions are those for 1084 field observations.

TABLE 1—*Continued.*

Population attributes:
Average longevity=10,489·5 years/1000 animals=10·5 years
Average weight=5041 pounds (2205 kilograms)
Standing crop=5·379 elephants per square mile (0·0208 per hectare)=27,116 pounds per square mile=4·75 grams per square meter which at 1·5 kilocalories per gram live weight=7·1 kilocalories per square meter.
Rate of tissue production=2,552,875 pounds/10,487·5 animal years=243·4 pounds (110·4 kilograms) per elephant per year. At 5·379 animals per square mile=1309·2 pounds of protoplasm per square mile (0·229 grams per square meter) produced annually. At 1·5 kilocalories per gram, net production is 0·34 kilograms per square meter per year.

meter per year yields the net production rate per unit area. Examples of these computations are given in Table 2.1.

**Trophic efficiency.** A population of herbivores does not exist as a separate entity but is an integral part of the biotic community, comprising a portion of a food web. The community of organisms forms an open system into which energy enters as sunlight and from which energy leaves as exported organic matter or as heat. Solar energy is converted by green plants, through the process of photosynthesis, into chemical potential energy. Much of the energy available in plant protoplasm is consumed by herbivores and later transferred to carnivores or to other organisms of advanced trophic levels.

Of particular interest is the ability of an organism to utilize its environment and, in the far view, to maintain its abundance for a prolonged period. To be highly efficient, the species population should not only be able to utilize its resources fully and to counteract its enemies effectively but it should be capable of modifying the environment in ways which are favorable to its sustained welfare.

The efficiency of a population of large herbivores in converting its forage resources to animal meat is an important aspect of energetics investigations. In meeting the study objectives, the minimum data necessary are the quantities of forage produced, consumed, defecated, and utilized in the maintenance and growth of a standing crop of herbivores (Table 2.2). The trophi cefficiency of a population may be expressed through the absolute values of production per square meter per year and by comparative ratios of several sorts.

As evidenced by the data of Table 2.2, the ability of ungulate species to utilize the available forage and to convert it to herbivore protoplasm varies considerably by species. An elephant population, for example, exists as a high biomass standing crop. On equal areas of their native ranges, the elephant biomass supported is more than five times greater than for the deer population, yet the rate of total protoplasmic growth is only about half as much.

The energy value of food consumed by an elephant population is not much higher than that eaten by the considerably lower standing crop of deer. Much food passes through the elephant in relatively undigested form, however, so that the proportion of food assimilated (growth plus maintenance items, Table 2) by an equivalent weight of elephants is less than half that assimilated by the deer. Per unit of standing crop, the elephant population eats only one-fourth as much and assimilates and adds growth at only about one-tenth the rate of deer. The proportion of assimilated energy which goes into growth is nearly the same in the two species.

In comparison with cattle, the rate of meat production for the wild species on their coarse diets is low. These data for cattle are idealized, however, and relate to good range conditions, ideal grazing habits, and optimum efficiency

TABLE 2.2 Comparative energy relations among populations of large mammals (average data in kilocalories per square meter per year).

| | *Odocoileus virginianus*[1] | *Loxodonta africanus* | *Bos taurus*[2] |
|---|---|---|---|
| *Hoofed-animal productivity:* | | | |
| Standing Crop | 1·3 | 7·1 | 7·5 |
| Available food | — | 747·0 | 28.6 |
| Food consumed | 52·6 | 71·6 | 14·3 |
| Feces | 12·5 | 48·3 | — |
| Growth[3] | 0·64 | 0·34 | 0·86 |
| Maintenance[3] | 39·5 | 23·0 | — |
| *Herbivore efficiency ratios:* | | | |
| Growth/Standing crop | 0·5 | ·048 | 0·115 |
| Assimilated energy[3]/Standing crop | 33·9 | 3·3 | — |
| Food consumed/standing crop | 41·4 | 10·1 | 1·9 |
| Growth/Food consumed | ·012 | ·005 | ·060 |
| Assimilated energy[3]/Food consumed | 0·8 | 0·3 | — |

[1] Data from Davis and Golley (1963: 261) for whitetail deer in Michigan, U.S.A.
[2] Data from Petrides and Swank (1966) for elephants in Uganda, East Africa, and for beef cattle under idealized range conditions, western U.S.A.
[3] Energy flow=assimilated energy=food consumed less defecation=gross production= growth plus maintenance.

in forage conversion. For a given caloric intake on their preferred ranges, steers produce twelve times as much growth as elephants and five times as much as deer.

Summaries of energy relations and trophic efficiencies, such as those of Table 2, are the fundamental data of productivity studies. But, the complete ecological roles of herbivores should not be overlooked. Elephants, for instance, benefit other herbivores by pushing over trees and by digging water-holes in dry riverbeds. They have esthetic and scientific values which should not be neglected in appraising their complete worth to man (Petrides and Swank, 1966). Similarly, other species have cultural, scientific, recreational, and economic attributes which are in addition to any peculiar characteristics of productivity which they may possess. Energy-flow characteristics comprise basic information, however, and should be determined for all important species. This should be accomplished while natural populations still exist in relatively undisturbed environments. They will provide standards against which to measure man's capabilities to influence the productivities of organisms.

### References

BENEDICT, FRANCIS G. (1936) *Physiology of the Elephant*. Carnegie Inst. Wash. Publ. 474, 302 p.

BRODY S. (1945) *Bioenergetics and Growth*. New York. 1023 p.

DAVIS, DAVID E. & GOLLEY FRANK B. (1963). *Principles in Mammalogy*. New York, Reinhold Corporation. 335 p.

GRAHAM N.M. (1964). Energy costs of feeding activities and energy expenditure of grazing sheep. *Aust. J. Ag. Res.* **15**, 969–973.

KLEIBER M. (1961). *The Fire of Life*. Wiley & Sons, New York, 454 p.

MAUTZ, WILLIAM W. & PETRIDES, GEORGE A. (1967). The usefulness of chromium-51 in digestive studies of the white-tailed deer. Proc. 32nd N. A. Wildl. Conf., 420–429.

PETRIDES, GEORGE A. (1967). Use of chromium-51 in the determination of energy flow characteristics in mammals. Symposium on *Recent Advances in Tropical Ecology*. Varananasi, India (in press).

PETRIDES, GEORGE A. & SWANK, WENDELL G. (1965). Population densities and range carrying capacity for large mammals in Queen Elizabeth National Park, Uganda. *Zoologica Africana* **1**, 209–225.

PETRIDES, GEORGE A. & SWANK, WENDELL G. (1966). Estimating the productivity and energy relations of an African elephant population. *Proc. Ninth International Grasslands Congress*, Sao Paulo, *Brazil*. January, 1965, 831–842.

# Section III
# Components of Energy Flow

# 3

# Numbers and Densities

A T Bergerud

Accurate population estimates are the essential first step in calculating the biomass and productivity of large herbivores. Factors affecting abundance cannot be evaluated unless population changes can be detected. The purpose of this chapter is to briefly discuss basic considerations and methodology in estimating ungulate populations. Emphasis is placed on population estimates rather than trend statistics.

**General considerations.** Davis (1963) lists two basic assumptions in estimating animal populations: (1) 'mortality and recruitment during the period when data are collected are negligible or the estimates are corrected for these effects . . . (2) all members of the population have an equal (or known) probability of being counted.'

At the outset of planning one must have a clear idea of the objectives and scope of the productivity study. It is assumed that only one ungulate species is involved. The better one knows his species the more apt he is to design and secure a satisfactory population or density estimate. However, many ecological investigations must of necessity commence with a population census. Is the species to be censused an open or closed population unit? Are there arbitrary population limits posed by political boundaries? An accurate assessment of a portion of a population subject to periodic ingress and egress will have little meaning in population dynamics evaluations unless movement can be assessed.

The abundance of sedentary ungulates can be expressed as animals per unit area (density), while a total estimate of mobile and gregarious species would probably be more meaningful.

Total or complete counts are desirable since they are not plagued with sampling errors. Many free-ranging ungulate populations will have to be estimated from sample counts of representative portions of the populations. If sampling counts are anticipated, it is wise to consult a biometrician for assistance in determining the distribution, number and size of sampling units.

It would be advisable at the onset to design the censusing technique so that the various components of variability can be distinguished and enumerated perhaps by an analysis of variance (F-test), *see* Erickson and Siniff (1963). A marriage of ecological and statistical know-how will provide better results.

**Minimum standards.** The literature abounds with methods and unverified results of big game counts (*see* Hazzard, 1958). Census methodology has now evolved to the stage where minimum standards are required before census results are accepted. The following minimum census stands are proposed: (1) that the accuracy of the method be verified with a known population, or (2) that two completely independent census methods give similar results. Further, the estimated variance of the observed mean should be given or 95% confidence limits provided. Estimates of variance values are, of course, of value even if the design is systematic rather than random (Greig-Smith, 1957). Sampling should be continued until the true observed mean can be estimated with a 95% confidence interval of 20% of the mean, i.e. within 10% either way with only a 5% chance of being in error (*see* Cockran, 1963). Remember, confidence limits are valid only for *observed* values. If the basic data is in error, statistical window-dressing will not alter the fact.

An example of the testing of a technique vs. a known population is provided by Dasmann and Mossman (1962). They compared the results of a road strip census of several African ungulates with the actual population within a one square mile enclosure. For other examples, *see* Robinette *et al.*, 1954 and 1956.

An example of point (2), verification of results from two independent methods, is given by Talbot and Stewart (1964). An aerial photography count of a herd of wildebeests (*Gorgon taurinus*) indicated 197,411 animals while an aerial census gave 197,000—200,000 animals. Hazzard (1958) provides examples of aerial vs. ground counts. Unless the results of the two census methods agree, one should not conclude that the highest figure is accurate *per se*.

**Distribution patterns.** If direct sampling is anticipated, the technique will depend on the distribution pattern of the individual animals and aggregations. Individual animals and/or aggregations might be quite randomly distributed or possibly clumped. Distribution patterns can be expected to vary with habitat, season densities, composition, time of day, etc. An initial step might be to completely count and plot a portion of the population in a sample area to investigate the spacing between individuals and aggregations.

For valid results the sampling unit should generally be larger than the population pattern.

Such an initial plotting might show that the distribution conforms to one of several mathematical dispersal models. If the distribution conforms to a mathematical model, the results will be more easily handled by routine statistical methods and likely the sampling effort can be held to a minimum.

If the individuals or groups were randomly dispersed, a frequency histogram of densities (density classes on the abscissa, and number of samples on the ordinate) would be skewed to the right and would fit a Poisson distribution. A more frequent distribution is the negative binomial in which a greater proportion of samples contain less and more animal numbers than a random series. Macfadyen (1963) provides a discussion of distributions and several references for testing a set of data to determine what type of distribution they fit.

**Systematic vs. random sampling.** A problem faced by all investigators contemplating sampling is whether to locate samples in a random or systematic manner. In random sampling the location of each unit is independently determined, while in systematic sampling the location of each sample is contingent on the placement of the initial plot. Macfadyen (1963) lists the following three references that might help in such a decision: Flemming and Baker, 1936; Yates, 1949; and Greenberg, 1951.

Randomized sampling has the distinct advantage that an estimation of the precision of the mean can be secured and confidence limits calculated between estimates from different habitats, years and populations. However, random sampling is frequently impractical. Frequently more work is involved in moving between random plots than travelling systematic routes. Systematic sampling can provide more representative results per sampling intensity if the animals are clumped (Bergerud 1963).

An example of these relationships is shown in Figure 3.1. The winter distribution pattern of moose (*Alces alces*) and caribou (*Rangifer tarandus*) are compared based on census data from Newfoundland. From the Figure it is obvious that moose are more homogeneously distributed than the caribou. The very marked clumping of caribou into companies (Fig. 3.1) suggests that systematic transect sampling might be more suitable than randomized blocks, while the more uniformly dispersed moose possibly should be counted by block sampling. Calculation of densities for both species based on systematic vs. random sampling verified these premises (Table 3.1).

**Reduction of variance.** The value of randomization is reduced if variance between samples is excessive. Broad confidence limits preclude distinguishing the population differences required for management. Variance can usually be reduced by: (1) stratification of the population by density classes, (2) sampling when the animals are most homogeneously distributed, and (3) increasing the size, number and linearity of the sampling units.

*Stratification.* Individual ungulates will not be randomly distributed, but frequently aggregations will be homogeneously dispersed. Populations sampled at such times could be computed by multiplying the number of aggregations per unit area by the average ungulates per group. In Figure 3.1 some moose are aggregated in herds of 2 to 9 animals. Moose aggregations thus are more evenly dispersed than individual animals. However, in Figure 3.1 there are patches of higher density (moose yards) in the upper and right and left corners of the block imposed on a more general distribution. Actually if the area of the sample were enlarged, it would be evident that the entire block is a high density patch compared to surrounding densities.

It is clear that if density areas are stratified, homogeneity will be enhanced and overall sampling variance reduced. In clumped distributions, variance increases with density. Stratification permits proportional or optimal allocation of sampling effort to strata. Siniff and Skoog (1964) allocated sampling effort for 6 caribou strata on the basis of preliminary population estimates. Their stratified optimal sampling allocation reduced variance by more than half over that of simple random sampling.

Some subjectivity cannot be avoided in defining strata boundaries. If the homogeneity is in doubt, study separately. Stratification not only reduces variation but results in a more even distribution of blocks as well as reducing the physical difficulties of moving between samples.

*Homogeneity of dispersion.* Animal clumping results from social interactions and habitat preference. If the factors affecting distribution are minor or of proportional intensity, homogeneity is enhanced. The more uniformly the habitat fulfills the species requirements, the more uniform the distribution. The more specific a species' needs the more clumped the distribution.

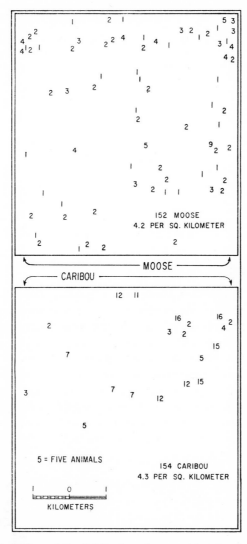

Figure 3.1. Comparison of the distribution of moose (above) and caribou (below) in two census blocks in Newfoundland that received 100% aerial coverage. The moose are more homogeneously distributed than the caribou because of a smaller aggregation size and uniform dispersal of aggregation.

*Chapter 3*

TABLE 3.1. Comparison of the accuracy of systematic transets vs. randomized block census techniques in censusing the moose and caribou distributions shown in Figure 3.1.

| Per cent Aerial Coverage | Deviations in Animals per sq. kilometer from actual density | | | |
|---|---|---|---|---|
| | Moose | | Caribou | |
| | Systematic | Random | Systematic | Random |
| 8 | − 2·5 | − 2·2 | +2·7 | +11·0 |
| 17 | − 0·9 | − 0·2 | +1·5 | − 4·3 |
| 25 | − 1·4 | +1·7 | − 0·1 | +1·5 |
| 33 | +1·8 | +1·0 | − 0·6 | +1·4 |
| 42 | − 1·2 | − 0·9 | +0·2 | +1·1 |
| 50 | +1·2 | +0·5 | +1·5 | +0·5 |

$$F = \frac{\text{Variance systematic}}{\text{Variance random}} = 1 \cdot 54 \qquad F = \frac{\text{Variance random}}{\text{Variance systematic}} = 11 \cdot 89$$

$$F\alpha = \cdot 025 = 7 \cdot 15$$

Social ungulates are most homogeneously distributed when the aggregation size is minimal. In the Northern Hemisphere most ungulates show a contagious (aggregated) distribution in the winter when food supplies are limited or restricted by snow cover. In the more southern latitudes, desert or steppe species are less dispersed during the dry season.

An annual cycle of gregariousness is to be expected for most species. In Newfoundland caribou the annual cycle results in reduced aggregations in the summer and precalving periods. However, counts cannot be conducted in the summer because many of the animals are under the tree canopy. In the precalving period the animals are scattered and contrast sharply with the vegetative background.

*Quadrat size and number.* A quadrat sample should deviate as little as possible from the true population parameters of variance and mean. An efficient quadrat size will have a small dispersion as measured by its variance (Greig-Smith, 1957). There should neither be an excessive number of plots with no individuals or aggregations (too small) or many plots with excessive numbers without intermediate densities (too large). In general, as a population becomes more clumped, larger and more samples are required. Greig-Smith (1957) states: 'As quadrat size increases and approaches the size of patches, variance relative to the mean will rise sharply. If the patches are regular, it will then fall off again, ultimately reaching, or even falling below, the mean. If, however, the patches are themselves randomly or contagiously distributed, the high variance will be maintained.'

If the population (individuals and/or aggregations) is randomly distributed, the size of the quadrat will not affect the variance since the variance of a Poisson series is equal to its mean; hence the variance of a single quadrat is equal to the units counted in the quadrat. Thus the size of the quadrat should be sufficient to give as random a distribution as feasible.

One procedure to follow is to select various quadrat sizes from the sample population and test for interspersion and magnitude of variance as measured by the percentage of the standard error of the mean ($S\bar{x}/\bar{x}$).

Measurements can be taken between nearest adjacent aggregations or random pairs and interspersion calculated. The formula used by Clark and Evans (1954) is:

$$R = \frac{\bar{r}A}{\bar{r}E} \quad \text{where:} \quad \bar{r}A = \frac{\Sigma r}{N} \quad \text{and} \quad \bar{r}E = \frac{1}{2\sqrt{P}}$$

Where:

R = the index of dispersion

N = the number of measurements taken between nearest aggregations

$r$ = the distance between nearest aggregations

P = the density of the observed distribution expressed in units used in $r$.

In a random distribution R = 1, values less than one indicate a clumped distribution, and greater than one a uniform distribution.

Student's and chi square tables can be used to determine significant departures from randomness (Greig-Smith, 1957).

$$t = \frac{(S^2/\bar{x}) - 1}{\sqrt{2/n-1}} \quad \text{and} \quad X^2 = (S^2/\bar{x})\,(n-1)$$

Where:

$S^2$ equals the sample variance

$n$  number of quadrats

$\bar{x}$  mean number per quadrat

The sampling precision of quadrats of 1, 2, 3 and 4 square kilometers for the moose aggregations in Figure 3.1 is given in Table 3.2. A quadrat of 2 square kilometers had the most random index of interspersion ($S^2/\bar{x}$) (1·03 vs. 1·00 for random) and the lowest standard error relative to the mean (0·20) of the four sizes tested.

In selecting the quadrat size, consideration must also be given to edge-effects. In small quadrats the ratio of edge to area increases resulting in a proportional greater segment of the population on the periphery of the quadrat. Unless the boundaries can be readily identified, arbitrary decisions on whether an aggregation is in or out will have to be made. The tendency is commonly to include rather than exclude. However, this bias may not be too serious in big game aerial counts since other visual biases tend to under-estimate the animals present.

Lastly, the quadrat size should allow a reasonable *normal* symmetry of the distribution curve based on the aggregations per sample plotted on the

TABLE 3.2. Determination of the proper siqe quadrat for sampling the moose population shown in Figure 3.1 based on samples of 8 quadrats.

| Kilo-meters per quadrat | Mean groups per quadrat ($\bar{x}$) | Sample Vari-ance ($S^2$) | Stan-dard error ($S\bar{x}$) | Ratio of S.E. to mean ($Sx/x$) | Index of inter-spersion ($S^2/x$) | Chi sq. value | Student t value |
|---|---|---|---|---|---|---|---|
| 1 | 1·13 | 1·27 | 0·40 | 0·35 | 1·13 | 7·91 | 0·24 |
| 2 | 3·25 | 3·36 | 0·65 | 0·20 | 1·03 | 7·21 | 0·06 |
| 3 | 4·00 | 8·57 | 1·04 | 0·26 | 2·14 | 14·98 | 2·13 |
| 4 | 9·50 | 73·14 | 3·02 | 0·32 | 7·70 | 53·90 | 12·54 |

ordinate and number or percentage of samples plotted on the abscissa (Greig-Smith, 1957). This is necessary since tests of significance between means usually require that means be normally distributed, and they may not be if the parent distribution is highly asymmetrical. If the mean is small and the distribution pattern is Poisson and skewed, it may be necessary to transform the sample values to their square roots for t-tests. A more satis-factory transformation if the mean is less than 10 is $\sqrt{X} + 0·5$ (Greig-Smith, 1957).

The question of quadrat shape can also be settled with a plot study com-paring variance between quadrats with varying length-width ratios. A rectangular block commonly provides more representative results than a square, especially if it is located at right angles to the contours (implying perhaps the usefulness of stratification).

The question of number of replications, other than practical considerations, is one of deciding what confidence limits are acceptable to accomplish the purpose of the study. Many small quadrats will be more representative than fewer larger plots covering the same portion of the ground. Variance tends to be reduced because there are more degrees of freedom and often it is possible to group plots into strata if adequate samples are available to partition sources of variation.

The minimum number of samples can be ascertained by cumulating replications and determining the point at which increased sampling does not materially alter the estimated density (Figure 3.2).

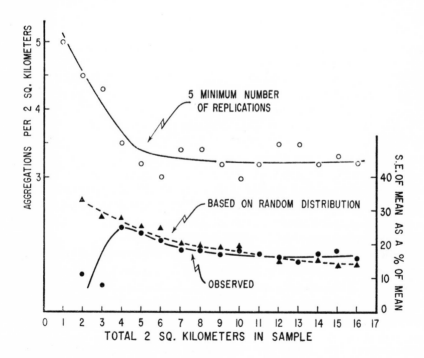

Figure 3.2. The use of cumulative samples in a moose census to determine the minimum number of replications.

If the minimum acceptable variance is known, variance can be calculated as sampling progresses and sampling can cease when the desired results are secured. If the distribution is random, the equality of the variance and mean implies that the standard error of the mean can be approximated by

$$\frac{\sqrt{X}}{n}$$ (X = total aggregations, $n$ = number of quadrats).

If it is desired to demonstrate a difference at the 5% level between two populations with respective means of 2 and 3 aggregations per kilometer, with some specified level of confidence (P), the sample size required per population is given by Steel and Torrie (1960) as:

$$n = \frac{2(t_0 + t_1)^2 s^2}{d^2}$$

Where:

    $t_0$ is the $t$ value for $2n - 1$ degrees of freedom at the specified level of significance

    $t_1$ is the $t$ value for $2n - 1$ degrees of freedom at the level of probability $2(1 - P)$

    $s^2$ is the estimated variance

    $d^2$ is the difference to be demonstrated

In a crude way, if it is desired to show no overlap at the 95% level between two populations of respectively 2 and 3 mean aggregations per kilometer, then a minimal standard of error of the mean number of aggregations of 0·24 is required for the more dense population; 95% confidence limits equal the $\bar{x} \pm S_{\bar{x}} t_{05}$ ($t_{05}$ equals approximately 2). The means $\pm$ 95% confidence limits would be approximated by $2 \pm 0·48$ and $3 \pm 0·48$.

Snedecor (1956) suggests as a very general rule that 20—40 replications are desirable. Less than 10 in most clumped distributions would certainly be inadequate. The decision on the number of samples must be decided in conjunction with quadrat size. Macfadyen (1963) provides several references on the subject.

**Direct counts.** The obvious and the most accurate method of determining the size of an ungulate population is to directly enumerate the individuals by complete and sample counts from the air or ground.

*Aerial counts.* Aerial counts of ungulates have been used most extensively in flat open habitats—grassland, desert, and tundra biomes. Aerial census is the only practical means of counting largei animals in wilderness terrain (Fuller, 1950). A disadvantage of aerial surveys is that they consistently underestimate densities—biases are seldom compensating, some animals are either overlooked or not in view at the time of the count. The proportion not counted is not constant, varying with the time of day, season of the year, and a multitude of factors. This variability can only be assessed with repeat counts. The efficiency of the specific aerial methods relied on must be tested. For instance, a herd could be counted prior to migration and the results compared to a track count or ground count as the animals passed in migration (Manuel, 1965).

In aerial counts the animals should always be plotted on large-scale maps. Not only does plotting assist materially in knowing what has been counted, but it assists in evaluating movement during the survey. Plotted aggregations can be used to evaluate accuracy and visualize the distribution and dispersion pattern. As the distribution pattern unfolds, the aerial coverage design may have to be altered; frequently flying hours can be saved after herd boundaries become discernable. Stratification requires knowledge on dispersion.

Specific considerations in planning an aerial survey are: (1) type of plane to use; (2) season and time of count; (3) altitude to count from; (4) observer efficiency; and (5) selection of complete, systematic, or random search designs.

The proper aircraft and pilot can 'make' a survey. Nothing is more distressing than 'whipping' across your population at speeds too great to accurately count with a pilot who is unfamiliar with game work and would rather be someplace else. Generally the slower the plane and the greater its visibility, the better. *See* Riordan (1948) and Saugstad (1942) for pilot and plane qualifications. In transect flying it is desirable to have a plane that can seat 2 or 3 passengers; a navigator can then record and plot observations allowing observers to devote all their time to scanning the ground. An intercommunication system is of value.

Animals should contrast well with their background. The more uniform the colour and physiography of the background, the better. Winter counts are enhanced by fresh snow advertising tracks. Sunny days provide sharper images and more contrasting shadows. Standing or moving animals are easier to recognize than lying or stationary animals. Oblique sun rays reflect better from animals than overhead lighting.

Flight altitudes depend on the ease of distinguishing animals. The higher the altitude, the more time is available to scan the ground and the greater the ground coverage. Eye fatigue is also less at moderate rather than low altitudes. Altitude depends also on ground strip width—a ratio of 1:1 or 1:2 between altitude and strip width is frequent.

Observer efficiency varies vastly between persons. Observers most familiar with the species should be used, and the efficiency of all observers should be tested against veterans of known capabilities. It is hard work searching for animals—the eyes must keep continually moving, systematically covering the area passing. Frequent rests at 2—4 hours improve observer efficiency. More animals will be missed if observers must record or segregate the animals seen. The longer the eyes remain locked with one aggregation, the more likely another group will pass unnoticed. More animals are missed from failing to note an aggregation than in undercounting a group (Bergerud, 1963). On transect runs at speeds greater than 80 m.p.h. it is difficult to count groups in excess of 30 animals. Aerial photographs can be taken of such groups and the animals tallied later.

*Complete counts.* Complete aerial counts cannot be attempted without accurate maps. It is necessary to know what animals have to be counted; this can be accomplished by subdividing the census area into smaller units and plotting groups counted. There is a natural tendency to miss smaller aggregations. Unfortunately there is no way to pick up these less conspicuous groups, since the plane must remain high so as to not disturb and mingle groups. Again, the identity of groups is lost at low altitudes. Complete counts must be accomplished quickly when the population is stationary. At times more than one plane can be used to advantage. Helicopters, because of their reduced speeds, allow more time to count and plot groups.

**Systematic transect counts.** In systematic aerial transect counts, flight lines should be designed to run at right angles to contours and cross the herd at its narrow width (Bergerud, 1963). The desired interval between transects can be partially gauged by comparing the spacing of herd areas along the transects. If, for instance, animals are continuously sighted for a distance of 8 kilometers, the herd area, if square, is 64 kilometers. Such a herd might be entirely overlooked with lines spaced at 10 kilometers intervals.

The width of transects can be ascertained by trigonometric functions based on angles-of-sight (Banfield *et al.*, 1955). A second method is to use predetermined strip widths delimited by marks on the struts placed by sighting in the air at marker flags on the ground set at known distances from the plane's flight path. Allowance must be made for a blind spot beneath the plane unless one observer scans forward.

A large number of inaccuracies can creep into a strip census. The herd should be stationary during the survey; animals must not be flushed into adjacent flight lines. Less time is available to search the ground than in block flying—the plane should not leave the flight line in attempts to count large groups. Such circling tends to group animals, some of which were probably not in the census strip originally. There is a natural tendency for observers to look outside the prescribed strip thereby missing animals within the transect. Counts may have to be increased 20% or more to compensate for missed animals. At times it is difficult to maintain the necessary altitude to give the desired transect width. Some pilots have an unconscious habit of deviating from predetermined flight lines and edging towards concentrations or, again, tipping the plane so that they can see better.

The accuracy of herd boundaries will depend on the spacing between lines. Manuel (1965) reported the following differences in one herd with different intensities of coverage:

| Distance between lines | | Area of herd | |
| --- | --- | --- | --- |
| Miles | Kilometers | Sq. miles | Sq. kilometers |
| 1 | 1·6 | 189 | 304 |
| 2 | 3·2 | 180 | 290 |
| 3 | 4·8 | 135 | 217 |
| 4 | 6·4 | 198 | 319 |
| 5 | 8·0 | 234 | 377 |
| 10 | 16·1 | 356 | 573 |

**Random block counts.** Randomized aerial block counts are ideal for open-living populations that are homogeneously distributed (Figure 3.1, top). The first step is to delimit the census area. Soil fertility classes or vegetative formation types might be used. The area can then be covered on preliminary systematic flight lines and animal locations noted. From the systematic lines the area is then stratified into density classes. Each stratum is then gridded into quadrats and the quadrats numbered. Quadrats to be censused are selected from a table of random numbers or by drawing numbers.

The method of censusing each quadrat will depend on its size. If the quadrats are larger than a few square kilometers, they will have to be subdivided. Usually the perimeter of the block is flown first to gain familiarization with boundaries. Small quadrats can be covered by circling into the center and then reversing the pattern–observer effort is concentrated on one side of the plane at a time. This method requires that accurate maps be available in order to recognize ground boundaries.

**Ground counts.** Ground counts will have to be relied on in forested habitat. Ungulates can be counted wherever they congregate: along rivers, at water holes, salt licks, wallows, etc. The difficulty is determining what proportion of the animals present are being recorded. To overcome this problem complete counts, drive or strip surveys can be undertaken.

*Complete count.* A complete count of a population is feasible only for intensively studied discrete populations. Andersen (1953) completely counted a small population of roe deer by killing the entire herd. The population was thought to contain 70 animals; 213 were collected and 4 or 5 escaped. H. Cummings (pers. comm.) randomly searched a Scottish forest for roe deer (*Capreolus capreolus*) stags. Antler diagrams were made of each stag observed in order to recognize the individual on subsequent sightings. He totalled his observations of identified stags and compared it to a Poisson distribution to estimate the number of stags that he may never have seen. (Table 3.3). This technique was successful because the stags were territorial and ingress-egress problems did not result.

A complete count of the California-Oregon deer herd was accomplished by counting tracks along a road that intersected the migration route. Tracks had to be brushed over each day to distinguish new tracks (Interstate Deer Herd Committee, 1950).

A similar technique was used in Newfoundland to count migrating caribou tracks in snow crossing logging roads (Manuel, 1965). Frequently the animals approached the road single file but spread out on the shoulder of the road permitting a count. Again the animals had to cross two roads separated by a river. In swimming the river the animals became further separated permitting more accurate counts on the second road. Track counts like aerial surveys provide conservative estimates.

*Drive census.* 'A drive is a method of censusing animals usually applied . . . that a straight line of drivers move across a selected area and either force

TABLE 3.3. Observations of roebucks identified by antlers at Glen Dye in 1963 (data from Harold Cummings).

| Number of repeat observations | Number of roebucks in each class | Expected No. of roebucks each class | Contribution to Chi-square |
|---|---|---|---|
| 0 | — | 0·948 | |
| 1 | 3 ⎫ | 2·791 ⎫ | |
| 2 | 4 ⎬ | 4·108 ⎬ | 0·148 |
| 3 | 5 | 4·106 | 0·241 |
| 4 | 2 ⎫ | 2·967 ⎫ | |
| 5 | 0 | 1·783 | |
| 6 | 3 ⎬ | 0·857 ⎬ | 0·000 |
| 7 | 1 ⎭ | 0·361 ⎭ | |
| Total | 18 | 17·831 | 0·389 |

the animals back through the line or out between the counters stationed around the periphery' (Hazzard, 1958). Deer drives have been used extensively in North America when abundant manpower was available (Hosley *et al.*, 1936; McCain, 1939; Olson, 1942; Morse, 1943; O'Roke and Hamerstrom, 1948.)

Areas to be driven could be selected randomly or as representative. Often selection will be dictated by terrain. It is essential to see the animals as they leave the block or double back through the line of beaters. If the animals are reluctant to leave the block, it is well to drive a triangular block, so that drives become more closely spaced as they approach the finish line.

*Strip census.* The strip or ground transect method involves travel on the ground through representative habitat on predetermined systematic routes where all the animals seen are counted and measurements taken to their sight location. Line-of-sight measurements are used in the King census method; in the Webb method measurements are taken perpendicular from the transect to the flushing locale (Webb, 1942). In the Hayne (1949) method sight distances are divided into frequency belts. The formula of the King census is

$$P = \frac{AZ}{ZYX}$$

Where:

P = population
A = total area of study
Z = number flushed
Y = average flushing distance
X = length of line

Hayne (1949) has criticized the King census on the assumption that the average flushing distance observed is not a good estimate of the true flushing distance.

When strip census results have been compared with a known population or used to estimate the number of dead deer, the results have not been encouraging (Rasmussen and Doman, 1943, Robinette *et al.*, 1956). The technique should be used in conjunction with other methods until its accuracy has been proven. Teer's *et al.* (1965) recent paper indicates some of the statistical documentation involved in presenting a convincing case for the strip census.

**Estimates based on sex, age, or marked ratios.** Census estimates are frequently based on the ratio of marked to unmarked animals in the population or resultant changes in sex and age ratios following known losses of specific sex or age classes.

*Marked-unmarked ratios.* The Lincoln Index is the most widely used marked- unmarked ratio and is calculated from $t = M/(x/n)$ or $t = Mn/x$

Where:

$t$ = estimate of population
M = number of animals originally marked
$x$ = number of marked animals in sample
$n$ = number of animals in the sample

Robson and Regier (1964) provide a discussion of the sampling and estimation problems for these kinds of estimates, and a series of figures which relate *n*, *x*, and M to confidence limits on the eventual estimate of population size. Their treatment underlines the importance of considering statistical analyses before a working experiment is conducted, thus ensuring that the proposed program will achieve the required objectives with optimum allocation of resources.

Adams (1951) and Ricker (1958) list the following conditions that should hold if the Lincoln Index is to be valid:

1. that marked animals suffer the same mortality as the unmarked,
2. that the marked animals must not lose their marks,
3. the marked animals must be subject to the same sampling as the unmarked animals,
4. the marked animals must become randomly mixed with the unmarked ones,
5. that all marked animals are recognized and reported on recovery,
6. that there is only a negligible amount of recruitment to the population during the sampling period.

The Lincoln Index in the past has not received the attention it warrants in the census of big game. Many species lend themselves to capture with the dart-immobilization technique or could be marked from aircraft. For example, in an interval of one hour 26 moose were sprayed with yellow paint from a helicopter in Newfoundland.

A modification of the Lincoln Index is the Schnabel (1938) Index which can be based on daily observations while marking is in progress. Davis (1963) mentions several other methods (*see* Schumacher and Eschmeyer (1943) and Chapman (1951)).

Kabat *et al.* (1953) used both the Lincoln Index and Schnabel method to census white-tailed deer in the Flag deer yard. Their marked animals were 23 animals that were recognized by color, scars, limping characteristics, etc. In a period of 9 days they visited 6 stations two or more times and saw 730 deer and their 23 identified deer 43 times.

$$t = (23)(730)/43 = 390$$

In using the Schnabel procedure the following observations were necessary: total deer seen, marked deer available in sighting, and repeat observations of marked deer. By this method the population was estimated to contain 489 individuals.

Kabat's *et al.* (1953) reliance on identifiable animals illustrates a pertinent point; it is not always necessary to capture and tag animals for future identification. Many ungulates can be identified by antler and horn characteristics (cf. Linsdale and Tomich, 1953). Cervids with palmate antlers are especially suited to this technique. Davis (1963) illustrates several examples of how records can be accumulated of identifiable animals to arrive at total estimates.

**Sex and age ratios.** Changes in the sex ratios following a kill of known number of bucks and does have frequently been used in North America to estimate deer populations.

Petrides (1949) stated: 'Formulas based on ratios and kill figures may not be applicable to some species at present because of difficulties in obtaining the required accurate data. Theoretically, however, they are applicable to any species. Furthermore, they function as well on a state-wide basis as on a smaller area; and they can often be used to check one another.' Chapman and Murphy (1965) have devised several ways of estimating variance in these types of situations. Hazzard (1958) and Hanson (1963) review the various formulas and related literature.

An example of this approach for mule deer is provided by Kelker (1944):

Where:

$1.948$ = ratio of does per buck before hunting season
$3.346$ = ratio of does per buck after hunting season
$0.891$ = ratio of fawns per doe after hunting season
$11,550$ = kill of bucks
$9,825$ = kill of does
$B$ = number of bucks before hunt
$1.948B$ = number of does before hunt
$B—11,550$ = number of bucks after hunt
$3.346 (B—11,550)$ = number of does after hunt
$1.948B—9,825$ = number of does after hunt

Hence:

$3.346 (B—11,550) = 1.948B—9,825$ where $B = 20,631$
$20,631 - 11,550 = 9,081$ bucks after hunt
$3.346 \times 9,081 = 30,385$ does after hunt
$0.891 \times 30,385 = 27,073$ fawns after hunt

Total post hunting population = 66,539.

Davis (1963) describes how a change in the frequency of any animal sign that varies directly with numbers can be used in conjunction with a known removal to estimate the population. 'For example, the number of deer tracks per area $(T_1)$ . . . may be counted, a known number of deer $(n)$ removed and then the tracks $(T_2)$ counted again.' Here we have

$$\frac{T_1 - T_2}{n} = \frac{T_1}{N_1} = \frac{T_2}{N_2}$$

Where:

$$N_1 = \text{population before removal}$$
$$N_2 = \text{population after removal}$$

An important assumption is that the reduction in density will not affect the number of tracks per animal. Assumptions such as this must be tested.

In fisheries, attempts have been made to estimate populations on the basis of harvest and age structure data (Ricker, 1958). By this method annual catches of a given year-class are summed until the age class disappears from the catch. One of the principal problems of the technique is estimating losses from other than legal harvest (Eberhardt, 1960). This method has not received wide use in ungulate investigations because of the inability until recently of distinguishing annual cohorts. This method will take on new importance now that many ungulates can be aged by counting the annuali in their teeth cementum (*see* Eberhardt, 1960, for further details).

**Indices.** Indices refer to ungulate signs that vary proportionately with densities and can be counted and converted to animal units. However, all indices must be compared with known populations and specific conversion coefficients calculated for the particular species composition, time, and place. Seldom can these conversion factors be corrected for the numerous observational and environmental biases. Hence indices lend themselves best to trend statistics.

The pellet group count is the most common index found in big game literature. Two problems are determining the daily defecation rate of the species and the durability of the pellets. Further, pellet groups are not randomly scattered so that sampling problems can be expected. Michigan deer workers have probably tested and refined the pellet count to a higher level of sophistication than elsewhere (Eberhardt and Van Etten, 1956; Ryel, 1959; and Eberhardt, 1960); yet in a recent paper Van Etten and Bennett (1965) concluded that the method was of limited value since workers incorrectly aged pellets and found varying proportions of the groups. Riney (1951) and Hazzard (1958) provide review papers.

Other index counts include tracks, beds, shed antlers, and calls. Russian biologists have estimated populations from track encounters recorded on aerial or ground transects. (Formozov, 1932 and Semenov, 1963.)

### References

ADAMS L. (1951). Confidence limits for the Peterson or Lincoln index used in animal population studies. *J. Wildl. Mgmt.* **15**, 13–19.

ANDERSON J. (1953). Analysis of a Danish Roe-deer population. *Danish Rev. of Game Biol.* **2**, 127–155.

BANFIELD A.W.F., FLOOK D.R., KELSALL J.P. & LOUGHREY A.G. (1955). An aerial survey technique for northern big game. *Trans. N. Amer. Wildl. Conf.* **20**, 519–532.

BERGERUD A.T. (1963). Aerial winter census of caribou. *J. Wildl. Mgmt.* **27**, 438–449.

CHAPMAN D.F. (1951). Some properties of the hypergeometric distribution with application to zoological sample censuses. *Univ. Calif. Pub. Stat.* **1**, 131–160.

CHAPMAN D.G. & MURPHY G.I. (1965). Estimates of mortality and population from survey-removal methods. *Biometrics.* **21**, 921–935.

CLARK P.J. & EVANS F.C. (1954). Distance to nearest neighbours as a measure of spatial relationships in populations. *Ecol.* **35**, 445–451.

COCHRAN W.G. (1953). *Sampling Techniques.* John Wiley and Sons, Inc., New York. 330 p.

DASMANN R. & MOSSMANN A.S. (1962). Road strip counts for estimating numbers of African ungulates. *J. Wildl. Mgmt.* **26**, 101–104.

DAVIS D.E. (1963). Estimating the numbers of game populations. pp. 89–118 in *Wildlife Investigational Techniques,* ed. H.S. Mosby, The Wildlife Society. 419 p.

EBERHARDT L. (1960). Estimation of vital characteristics of Michigan deer herds. *Game Div. Rep. No.* 2282. Michigan Dept. Conservation, Lansing. 192 p.

EBERHARDT L. & VAN ETTEN R.C. Evaluation of the pellet group count as a deer census method. *J. Wildl. Mgmt.* **20**, 70–74.

ERICKSON A.W. & SINIFF D.B. (1963). A statistical evaluation of factors influencing aerial survey results on brown bears. *Trans. N. Amer. Wildl. and Nat. Resources Conf.* **28**, 394–409.

FLEMMING W.E. & BAKER F.E. (1936). A method for estimating populations of larva of the Japanese beetle in the field. *J. Agric. Res.* **53**, 319–331.

FORMOZOV A.N. (1932) Formula for taking a census of mammals by their tracks. *Zoologichestii Zhurnal.* **11**, 66–69.

FULLER W.A. (1950). Aerial census of bison in Wood Buffalo Park. *J. Wildl. Mgmt.* **14**, 445–451.

GREENBERG B.G. (1951). Why randomize. *Biometrika.* **7**, 309–332.

GRIEG-SMITH P. (1957). *Quantitative Plant Ecology.* Butterworths Scientific Publications, London. 198 p.

HANSON W.R. (1963). Calculation of productivity, survival, and abundance of selected vertebrates from sex and age ratios. *Wildl. Monog.* **9**, 1–60.

HAYNE D. (1949). An examination of the strip census method for estimating animal populations. *J. Wildl. Mgmt.* **13**, 145–157.

HAZZARD L.K. (1958). *A Review of Literature on Big Game Census Methods.* Colorado Game and Fish Dept. 76 p.

HOSLEY N.W., ASHMAN R.I., BRODDER W.E., DALKE P.D. & MOSS A.E. (1936). Forest wildlife census methods applicable to New England conditions. *J. For.* **34**, 467–471.

INTERSTATE DEER HERD COMMITTEE (1950). Fourth progress report on the cooperative study of the Interstate Deer Herd and its range. *Calif. Fish and Game.* **36**, 27–52.

KABAT C., COLLIAS N.E. & GUETTINGER R.E. (1953). Some winter habits of white-tailed deer and the development of census methods in the flag yard of Northern Wisconsin. *W.S. Conserv. Dept. Tech. Wildl. Bull.* **7**, 1–31.

KELKER G.H. (1944). Sex ratio equations and formulas for determining wildlife populations. *Proc. Utah Acad. Sci., Arts and Letters.* **19–20**, 189–198.

LINSDALE J.M. & TOMICH P.Q. (1953). *A Herd of Mule Deer.* Univ. of Calif. Press, Berkely and Los Angeles. 567 p.

MACFADYEN A. (1963). *Animal Ecology: aims and methods.* Sir Isaac Pitman and Sons Ltd., London. 344 p.

MANUEL F. (1965). *Newfoundland Big Game Aerial Census Techniques.* Dept. of Mines, Agric. and Res., Newfoundland. 10 p.

MCCAIN R. (1939). The development and use of game drives for determining whitetail deer populations on Allegheny National Forest. *Trans. N. Amer. Wildl. Conf.* **4**, 221–230.

MORSE M.A. (1943). Techniques for reducing man-power in the deer drive census. *J. Wildl. Mgmt.* **7**, 217–220.

OLSON H. (1942). Superior forest deer studies. *Conserv. Volunteer.* **3**, 52–56.

O'ROKE E.C. & HAMERSTROM JR. F.N. (1948). Productivity and yield of the George Reserve deer herd. *J. Wildl. Mgmt.* **12**, 78–86.

PETRIDES G.A. (1949). Viewpoints on the analysis of open season sex and age ratios. *Trans. N. Am. Wildl. Conf.* **14**, 391–410.

RASMUSSEN D.L. & DOMAN E.R. (1943). Census methods and their application in the management of mule deer. *Trans. N. Am. Wildl. Conf.* **8**, 369–379.

RICKER W.E. (1958). Handbook of computations for biological statistics of fish populations. *Fisheries Res. Board of Canada Bull.* **119**, 1–300p.

Riney T. (1957). The use of faeces counts in the studies of several free ranging mammals in New Zealand. *N.Z. For. Serv., Wellington, N.Z., N.Z.J. and Tech. Ser., B.* **38**, 507–532.

RIORDAN L.E. (1948). The sexing of deer and elk by airplane in Colorado, *Trans. N. Am. Wildl. Conf.* **13**, 409–430.

ROBINETTE W.L., JONES D.A., GASHWILER J.S. & ALDOUS C.M. (1954). Methods for censusing winter-lost deer. *Trans. N. Am. Wildl. Conf.* **19**, 511–525.

ROBINETTE W.L., JONES D.A., GASHWILER J.S. & Aldous C.M. (1956). Further analysis of methods for censusing winter-lost deer. *J. Wildl. Mgmt.* **20**, 75–78.

ROBSON D.S. & REGIER H.A. (1964). Sample size in Petersen mark-recapture experiments. *Trans. Am. Fish Soc.* **93**, 215–226.

RYEL L.A. (1959). Deer pellet-group surveys on an area of known herd size. *Twenty-first Midwest Wildl. Conf., Mich. Dept. of Conserv., Lansing Game Div. Rep. No.* 2252, 1–26.

SAUGSTAD S. (1942). Aerial census of big game in North Dakota. *Trans. N. Am. Wildl. Conf.* **7**, 343–356.

SEMENOV B.T. (1963). Censusing game animals in daily activity plots (martens and elks) or individual territories (otters). p. 27–28 in *Organization and Methods of Censuing. Terrestrial Vertebrate Faunal Resources.* ed. Y.A. Isakov. National Scientific Foundation, Washington, D.C. 104 p.

SCHNABEL Z.E. (1932). Estimation of the total fish population of a lake. *Am. Math. Monthly.* **45**, 348–352.

SCHUMACHER F.X. & ESCHMEYER R.W. (1943). The estimation of fish populations in lakes and ponds. *J. Tenn. Acad. Sc.* **18**, 228–249.

SINIFF D.B. & SKOOG R.O. (1964). Aerial censusing of caribou populations using stratified random sampling. *J. Wildl. Mgmt.* **28**, 391–401.

SNEDECOR G.W. (1956). *Statistical Methods.* Iowa St. College Press, Ames, Iowa. 534 p.

STEEL G.D. & TORRIE G.H. (1960). *Principles and Procedures of Statistics.* McGraw-Hill Book Co., New York. 481 p.

TALBOT L.M. & STEWART D.R.M. (1964). First wildlife census of the entire Serengeti-Mara region. *J. Wildl. Mgmt.* **28**, 815–827.

TEER J.G., THOMAS J.W. & WALKER E.A. (1965). Ecology and management of white-tailed deer in the Llano Basin of Texas. *Wildl. Monog.* **15**, 62 p.

VAN ETTEN R.C. & BENNETT, JR. C.L. (1965). Some sources of error in using pellet-group counts for censusing deer. *J. Wildl. Mgmt.* **29**, 723–729.

WEBB W.L. (1942). Notes on a method for censusing snowshoe hare populations. *J. Wildl. Mgmt.* **6**, 67–69.

YATES F. (1949). *Sampling Methods for Censuses and Surveys.* Griffin, London, 318 p.

# 4

# Categories and Age Determination

R C BIGALKE

A population usually contains a wide range of individuals of different ages and different sizes. In order to study the population we must determine its age structure and the rate of growth or decline. These two characteristics are of course interdependent, for individuals do not usually reproduce in infancy or old age and the age structure will obviously be reflected in the growth rate. Thus a population containing a high proportion of individuals too young or too old to reproduce, will increase at a much lower rate than one with a high proportion of reproducing individuals. The relative proportions of the two sexes must also be determined, for, to state the obvious, populations consisting entirely of either males or females will not increase. Thus a study of population structure as a whole involves the study of both its age and its sex structure—the determination of the proportions of males and females, and of such categories of immature, mature and aged individuals as can be distinguished. Time is an important factor. The structure of a population determined at a given instant will not remain constant for any length of time. The factors making for change are recruitment by the birth of new individuals, growth, and death. In addition, if the population under study is not an isolated one, immigration to or emigration from it may introduce additional complications.

A prerequisite for the study of population structure is the identification and definition of those age and sex categories which can be distinguished. The degree of refinement with which these categories of individuals can be differentiated will depend on a number of factors. The more important ones are:

(1) The material available for study—can only living animals be observed in the field or is it possible to handle and examine them at close quarters, or even to subject dead animals (or parts of them, e.g. skulls) to detailed examination in the laboratory? Usually only quite gross and obvious external differences can be recognized in the field. More refined criteria can be applied when sorting laboratory material.

(2) The morphological characteristics of the species.
(3) The reproductive pattern of the species—whether or not breeding is seasonal.
(4) Size and rate of growth.
(5) Behavioural characteristics.

It is very rarely possible to determine the structure of an entire population. The extermination of a roe deer herd in Denmark provided the opportunity for a classical study of this kind (Andersen, 1953.) However, few research workers can expect to have such exceptional good fortune.

Usually the structure of a sample must be determined and the results applied to the whole population. As with all sampling, it is necessary to ensure that the sample studied is random and unbiased and therefore representative of the population as a whole. It is important to realize that behaviour is often a major source of bias in sampling mammal populations. An investigator must therefore be familiar with at least the main features of the behaviour of the species he wishes to study before he can plan the collection of random samples.

The inexperienced may not only fail to see a large proportion of the animals present in a given area, but those seen and noted can tend to form a biased sample. The bias is likely to occur because of the complexity of the dispersion of individuals within a population, due to the difference in social organization of the sexes and age classes at different times of the year.

For example, males, either alone or in groups, often run separately from other classes for much of the year. In Scottish red deer most of the older stags are separated from the hinds for most of the year; in winter they tend to inhabit the lower ground and in summer the highest ground. In many antelopes, females have a greater flight distance (i.e. are shyer) when they have young at foot than at other times. Immatures of many species, e.g. eland, springbok, oryx, run together in segregated groups for some time before being incorporated into the adult groups.

Clearly in all these examples the pattern of behaviour must be known before it is possible to devise a system of sampling which will produce reliable results. However, in spite of these complicating factors the value of any count at any time of year will be greatly enhanced if the animals seen are classified into at least a few categories.

**Criteria for differentiating categories**

I. *Age criteria*

Field observers of large herbivores have used many different terms for the age classes they have distinguished. Examples are: new-born, infant, juvenile, yearling, young of the year, immature, sub-adult, young, adult, old, knobbers and spikers (yearling red deer stags), spike horn (2–3 and some 4-year-old bison). There are as yet no universally valid standardized definitions of most of these terms.

When carrying out a study, the terms used should be carefully defined at an early stage. If this is not done, it may necessitate the subsequent rejection of some of the field notes. The observer's skill in identifying age groups may improve as he gains experience. In the course of the study he may also discover new identifying features unknown to him at the outset. It is nevertheless important to aim at standardizing the categories employed as soon as possible.

(a) *Seasonal breeders.* The simplest case is that of herbivore species in which the annual period of parturition is short and distinct. A few random examples of species occurring in various parts of the world in which this is the case are: red deer in Scotland—90% of the year's calves born in a three-week period in May/June; saiga antelope in Russia—young born in a period of one month but most in a single week, usually the first week of May; vicuna in Peru—more than half of the young born in March.

In species which reproduce in this way, the young born in any one year constitute a uniform and easily recognizable group within the population. The length of time during which it is possible to distinguish each year class depends on the rate of growth and maturation of the species, and on the existence of morphological differences between the sexes. In a small, relatively fast-growing species such as impala, one season's young are quite readily distinguishable for the first six or nine months. Thereafter sexual differences become important. The hornless females, differing from adults only by their size, then become increasingly difficult to identify. In their second year of life, they no longer constitute a recognizable year class. On the other hand the males can be quite simply separated by the size and shape of the horns during both their first and second years. Only in the third year do they become indistinguishable from older animals.

In larger species which grow more slowly and mature later, more year classes are identifiable. For example, bison of both sexes can easily be classified during both the first and second years. Thereafter the categories become broader than year classes; two further categories can be distinguished in females and three in males.

In some species, parturition, although seasonal, is spread over a period of several months within which there may be a peak period. In such cases the young of a given year class will of course vary quite considerably in age and size, the youngest and smallest members of one approaching the size of the oldest and largest representatives of the following year class.

However many year classes it may be possible to recognize, two requirements are basic in studying any population of seasonally breeding herbivores.

(i) *Young of the year.* The first is to distinguish the young of the year (0–12 months old). The proportion of these animals gives a measure of reproductive success and hence of the rate of recruitment of new individuals into the population.

The time at which adult/young counts are made is important for two reasons. Firstly, species differ in the relationship between females and their newly born young. There are two main kinds of mother-young relationships in the bovids, and one can place most species in one or other of them. Fritz Walther (1960) has called these categories the 'followers' (Nachfolgetypus) and the 'hiders' (Abliegetypus).

In the followers, the young animal accompanies its mother constantly from the time it is physically able to do so. This is usually very shortly after birth. Examples of followers are chamois, blesbok and wildebeest. In hiders, the young animal lies up alone some distance away from the mother. She visits it periodically to nurse it. The period of hiding lasts for at least several weeks after birth. Examples of hiders are kudu, duiker and gazelles. There are patterns of behaviour intermediate between these two main types.

Accurate adult/young counts can be made much sooner after the end of the season of parturition in followers than in hiders.

The time at which the young are counted is also important because mortality is usually high during the first year of life. Counts made immediately after the season of birth reflect (approximately) the reproductive success. Thereafter they reflect natality minus deaths up to the time of the count.

If the same population is studied in successive years, counts must be made at the same time each year if they are to be comparable.

(ii) *Recruitment into the breeding population.* Important as it is to determine the proportion of young of the year, it is even more important to ascertain that of female animals which survive from the season of their birth until they become sexually mature and themselves produce offspring. This presupposes a knowledge of the age at which females first become pregnant and bear young. The proportion of surviving males is less important. In practice, reproduction is rarely prevented by a shortage of males. Speaking generally, they seldom have opportunities to mate with females of their own age, since older males tend to play a dominant role in the activities of courtship and copulation.

The age of sexual maturity (not necessarily synonymous with physical maturity) is a characteristic of each species, but it can vary greatly in response, e.g. to different levels of nutrition. Thus in New York State, more white-tailed deer does were found to reproduce in their first year on range of good quality than in areas where the habitat is less suitable for deer. Mammals usually tend to breed earlier in captivity than in the wild, so that data from captive specimens should be treated with circumspection.

In many small quick-maturing species, e.g. springbok, impala, most females are impregnated in their second year and bear their first young on the second anniversary of their own birth. Thus if the young of the year are counted as late in their first twelve months of life as possible, an approximation is obtained of the number of animals which will enter the breeding population.

Larger species that mature slowly pose more difficult problems. In red deer it is not possible to determine the sexually immature females in the field. In female bison a sub-adult nonbreeding age group cannot be recognized. However, because about one-third of the cows attain sexual maturity at two years of age and most of the others, probably, as three-year-olds, there is little need to separate such a group when making field counts.

In some species even young of the year cannot be easily sexed in the field. If, however, the males become distinguishable in their second year, e.g. red deer, they can be counted then as a separate category and the sex ratio of the previous year's calf crop can be determined in retrospect.

(b) *Non-seasonal breeders.* In species which breed throughout the year, or those which do so in certain parts of their range (usually in the tropics), it is difficult or impossible to recognize year classes in the field. A marsupial such as the red kangaroo poses particularly difficult problems. In addition to breeding continually, some females are accompanied simultaneously by two generations of young, one in the pouch and one at foot.

Annual counts are of little use. Classes must be set up, often of necessity characterized by arbitrary criteria of size (e.g. young as high as mother's belly), colour or of some morphological feature (e.g. length of horns). Counts must be repeated at frequent intervals. In this way the change in the relative proportions of each size class in the population can be determined. Absolute ages must then be assigned to the size classes by reference to animals of known age.

In a species as small as a duiker, only one non-adult class can be distinguished. In a large, slow growing species such as the elephant, 7 size classes are quite easily identifiable in the field.

## II. *Sex criteria*

In many large herbivores the sexes are quite dissimilar, usually because only males have horns or antlers, or because of marked sexual dimorphism in colour, body size or the development of tusks. Where sexing is difficult or even impossible, this limitation should be accepted. Time should not be wasted in fruitless efforts to ascertain sex ratios which, if obtained, would be of doubtful accuracy.

In cases where the sexes are morphologically similar, differences in behaviour may prove useful. In species where the males are (at least seasonally) territorial, e.g. some gazelles, the male accompanying a herd often tends to lag behind when the rest run off on being disturbed. A territorial male often adopts a conspicuous display posture, making no attempt at flight or concealment. The males of many antelope have a thick neck which is carried high when they run. Details of this kind often enable an experienced observer to determine the sex of animals at distances where other sexual characteristics are not discernible.

The technique of establishing the sex ratio of a crop of deer calves retrospectively, by counting the males when their antlers begin to grow in the second year, has already been mentioned. Walther has been able to determine the sex of newly born gazelles by observing the behaviour of mothers when they visit the hidden young. While suckling a lamb, the mother massages the excretory openings with her tongue, presumably partly to stimulate urination and defaecation, and partly to consume the faeces and urine. This is believed to reduce the chances of predators being attracted to the young animal by the smell of its excreta. By careful observation, it is possible to see whether the mother drinks urine from a vagina or from a penis, and hence to sex her offspring.

### III. *Age determination*

A great deal of effort has been devoted to the study of techniques for determining the age of large herbivores. More attention has of course been paid to domestic animals than to wild ones, the first known written reference to the subject being Xenophon's account, in 400 B.C., of tooth replacement in the horse. Good age determination criteria exist for wild species hunted for sport in Europe and North America, and for some marine mammals of economic importance. In general, little work has been done in other parts of the world.

Antlers, horns and teeth are the structures most often used to determine age. They are usually easily available for study and change in a regular fashion with increasing age, antlers and horns altering in shape and growing larger, while deciduous teeth are replaced by permanent ones which wear down. There are many other criteria which can be used to age large herbivores with greater or lesser degrees of precision.

It is important to remember that there may be some individual variation in the age at which changes in form and structures take place. Different populations of the same species may also differ significantly.

(a) *Teeth.* Because they are easily collected, lower jaws are often used alone in studying dental characters. If only live animals are available, tooth impressions may be taken from them.

(i) *Tooth eruption and replacement.* Most large herbivores have a set of deciduous or milk teeth consisting of incisors, canines and premolars, the number depending on the taxonomic group or species to which an animal belongs. In each species these teeth are replaced by permanent successors, and the permanent molars erupt through the gums in a characteristic and usually constant sequence. The ages at which the major events of replacement and eruption occur are, for the most part, constant within fairly narrow limits. However, exceptionally good or poor conditions for growth and development may result in unusual variability. Once the sequence has been established for a species by studying animals of known age, any individual can be aged by examination of the jaws. In most species, tooth replacement can be used to age animals up to about 4 years old.

(ii) *Tooth wear*. After the full set of permanent teeth has developed, ageing depends on the degree of wear on them. This is seldom a precise technique and regional differences, e.g. the hardness and abrasive qualities of the food plants, may cause wide variations in the rate of wear in different populations of the same species.

The degree of wear may be determined by measurement. The height of one or more cheek teeth above the gum line (crown height) is used alone, or a ratio is calculated between this and the width of the occlusal surface of the tooth (teeth). The width of the first incisor is a useful indicator of age in moose.

Other criteria of wear and hence of age which may be used are: changes in the pattern of enamel and dentine on the occlusal surfaces of the cheek teeth, changes in the shape and depth of the infundibula in the molars, and the progressive loss of various cusps. In some species it is possible to set up 'wear classes' based on the appearance of many or all of the teeth, which enable adult animals to be grouped into broad categories such as 'adult' or 'aged.'

(iii) *Growth layers*. Laws showed that in marine mammals seasonal influences are reflected in the structure of the teeth. This has subsequently been established for a number of terrestrial mammals. Internal growth layers are found in the dentine and cement of the teeth, and external ridges or annuli are often present on the roots.

Some species in which the cementum shows annuli which can be used for age determination are caribou, moose and bison. Some teeth may be more suitable than others for the application of this technique, e.g. the lower canine is the most satisfactory tooth in the bison. Layers of secondary dentine laid down in the pulp cavity of the first incisor are reliable indicators of age in red deer between about 3 and 10 years old.

(iv) *Tooth weight*. The weight of elephant tusks is related to age. There may be similar relationships between tooth weight and age in other species which have yet to be demonstrated.

(v) *Other criteria*. The following statistics have been found to be useful indicators of age in some species: the length of the row of premolars and molars, the length of the diastema between the canine and the first of the cheek teeth (premolar 2), the length of the lower jaw, the position of the molar tooth row relative to the zygomatic process (red kangaroo).

(b) *Antlers and horns.* The size and shape of antlers are generally of value only in distinguishing one or two younger age classes. The relationship between length and diameter of the antler pedicel has been used to set up a table for age determination in German red deer up to an advanced age.

The size and shape of horns has been used as an aid in ageing, e.g. bison and wildebeest. In grey duiker, horn height has been found to have an extremely limited value as a criterion of age. Winter check lines in the growth of horns can be counted in, for example, wild sheep, rocky mountain goat, and chamois, and provide a satisfactory method of age determination.

(c) *Bones and cartilages.* While mammals are growing, the long bones of the limbs lengthen by the deposition of bone in a cartilaginous area between the diaphysis (bony shaft) and the epiphysis (terminal bony cap). This epiphyseal cartilage is visible as a broad zone in young animals and as a line of decreasing width while it ossifies as growth gradually ceases. In an adult in which growth is complete, the line is no longer present and the epiphysis is said to be closed. The degree of closure of the epiphysis (4 degrees) in the bones of the limbs has been used to estimate the age of white-tailed deer up to 72 months.

Young growing animals may also be distinguished from old ones by the degree to which the sutures between the bones of the skull have closed.

The ossification of the thyroid cartilage in the larynx of the roe deer progresses in a regular manner. The form and extent of the ossification can be used as a criterion of age.

(d) *Body size.* Where a sufficiently large and representative body of statistics has been collected, it is often possible to identify particular measurements e.g. body weight, length, shoulder height, girth, which may be used as indicators of age. These are usually applicable mainly to younger age groups.

In adults the effects of environmental conditions often mask any possible correlations between body dimensions and age. However in some cases a measurement such as the length of the hind foot can be a reliable indicator of age. The pouch young of marsupials, e.g. red kangaroo, constitute a special case. They are easily measured and various body dimensions are useful and accurate age criteria.

In large, slow-growing species, e.g. elephant, the size of young animals relative to adults may be a useful field criterion. Seven size classes are distinguishable. Size classes should, as has been done for the elephant, be converted to absolute ages.

(e) *Colour*. In some species the young differ markedly from adults in body colour and pattern. The young of both sexes often resemble females, and the males then assume adult male colouration at a certain age. Examples are nyala, situtunga, sable antelope. The colour of such species can be used to determine age.

### Acknowledgments

I am most grateful to two colleagues who took much trouble over compiling reviews of categories and age criteria which have been of material assistance in the compilation of this chapter. Mr V.P.W. Lowe wrote up red deer in Scotland and Dr W.A. Fuller produced a detailed account of age and sex categories in Canadian big game. Dr. George Dunnet suggested sources of information in Australia and Mr H.J. Frith kindly sent me reprints of work done in that country on developing ageing criteria for kangaroos.

### References
(General works and recent papers)

ANSELL W.F.H. (1960). The breeding of some larger mammals in Northern Rhodesia. *Proc. Zool. Soc. Lond.* **134**, 251–274.

ARMSTRONG G.G. (1965). An examination of the cementum of the teeth of Bovidae with special reference to its use in age determination. M. Sc. thesis, Dept. of Zoology. Univ. of Alberta.

BANFIELD A.W.F. (1960). The use of caribou antler pedicels for age determination. *J. Wildl. Mgmt.* **24**, 99–102.

BANNIKOW A.G. (1963). *Die-Saiga-Antilope* (*Saiga tatarica L.*). Neue Brehm-Bücherei, Wittenberg Lutherstadt, A. Ziemsen Verlag.

BERGERUD A.T. (1961). Sex determination of caribou calves. *J. Wildl. Mgmt.* **25**, 205.

BERGERUD A.T. (1964). Relationship of mandible length to sex in Newfoundland caribou. *J. Wildl. Mgmt.* **28**, 54–56.

CHILD G. (1964). Growth and ageing criteria of impala, *Aepyceros melampus. Occ. Pap. Nat. Mus. S. Rhod.* **27B**, 128–135.

DENNEY R.N. (1958). Sex determination in dressed elk carcases. *Trans. N.A. Wildl. Conf.* **23**, 501–513.

ERZ W. (1964). Tooth eruption and replacement in Burchell's zebra, *Equus burchelli* Gray (1825). *Arnoldia* **1**, (22), 1–8.

FRITH H.J. & SHARMAN G.B. (1964). Breeding in wild populations of the red kangaroo, *Megaleia rufa. CSIRO Wildl. Res.* **9**, 86–114.

FULLER W.A. (1959). The horns and teeth as indicators of age in Bison. *J. Wildl. Mgmt.* **23**, 342–344.

GILBERT F.F. (1966). Ageing white-tailed deer by annuli in the cementum of the first incisor. *J. Wildl. Mgmt.* **30**, 200–203.

GEIST V. (1964). On the rutting behaviour of the mountain goat. *J. Mammal.* **45** (4), 551–568.

HABERMEHL K.J. (1961). *Die Altersbestimmung bei Haustieren, Pelztieren und beim jagd-baren Wild.* Berlin und Hamburg, Paul Parey.

KLINGEL H. (1965). Notes on tooth development and ageing criteria in the plains zebra *Equus quagga boehmi* Matschie. *E. Afr. Wildl. J.* **3**, 127–129.

KLINGEL H. & Y. (1966). Tooth development and age determination in the plains zebra (*Equus quagga boehmi* Matschie). *Der Zool. Garten (NF)* **33**, 34–54.

LAWS R.M. (1962). Age determination of pinnipeds with special reference to growth layers in the teeth. *Z. Saugetierk.* **27**, 12–146.

LAWS R.M. (1966). Age criteria for the African Elephant, *Loxodonta a. africana. E. Afr. Wildl. J.* **4**, 1–37.

LEWALL E.F. & COWAN I.McT. (1963). Age determination in black-tail deer by degree of ossification of the epiphyseal plate in the long bones. *Can. J. Zool.* **41**, 629–636.

LOW W.A. & COWAN I.McT. (1963). Age determination of deer by annular structure of dental cementum. *J. Wildl. Mgmt.* **27**, 466–470.

LOWE V.P.W. (1966). Observations on the dispersal of red deer on Rhum, in *Play, Explora-tion and Territory in Mammals*: Jewell P.A. & Loizos C. (eds.). Sympos. Zool. Soc. London. No. 18, 211–228.

McEWAN E.H. (1963). Seasonal annuli in the cementum of the teeth of barren-ground caribou. *Can. J. Zool.* **41**, 111–113.

MOSBY H.S. (ed.) (1963). *Wildlife Investigational Techniques.* 2nd ed. The Wildlife Society.

RANSOM A. BRIAN (1966). Determining age of white-tailed deer from layers in cementum of molars. *J. Wildl. Mgmt.* **30**, 197–199.

RINEY T. & CHILD G. (1964). Limitations of horn height as an index to ageing the common duiker (*Sylvicapra grimmia*) *Arnoldia.* **1** (1), 1–4.

SHARMAN G.B., FRITH H.J. & CALABY J.H. (1964). Growth of the pouch young, tooth eruption, and age determination in the red kangaroo, *Megaleia rufa. CSIRO Wildl. Res.* **9**, 20–49.

TAYLOR W.P. (1961). The white-tailed deer of North America, in *Ecology and Management of Wild Grazing Animals in Temperate Zones*, ed. F. Bourlière, 8th Technical Meeting, IUCN, Warszawa, 1960; reprinted from *La Terre et la Vie*, 1961, 3–157, 179–358.

TALBOT L.M. & TALBOT M.H. (1963). The Wildebeest in Western Masailand, East Africa, *Wildl. Monog.* No. 12.

WALTHER, FRITZ (1960). *Mit Horn und Huf.* Verlag Paul Parey.

# 5

# Measuring Quantity and Quality of the Diet of Large Herbivores

GEORGE M VAN DYNE

**Introduction.** Most of the methods and information concerning the quantity and quality of the diet of large herbivores has been derived in studies with domestic animals. This review places emphasis on those studies, for they possibly provide an upper limit to the precision and accuracy attainable in studies with wild herbivores. Primary emphasis is given to studies with grazing cattle and sheep rather than to investigations in metabolism crates or in feedlot trials. Methods used solely with wild animals are discussed and compared with those used only with domestic animals. Original literature is referenced throughout this review, but special attention will be given to review papers. Coverage is not intended to be exhaustive, but instead in many instances representative studies are cited.

Study of the quantity and quality of the diet of grazing animals has three basic components. *First*, estimates must be obtained of the chemical and botanical composition of the diet. In the current review emphasis is given to stomach analysis and fistula techniques. *Second*, estimates must be made of the digestibility of the diet. Herein, both field methods and laboratory methods will be discussed. It would appear that in future studies the field and laboratory techniques could be used in a complementary, double-sampling procedure. *Third*, to estimate total herbage intake of grazing animals a measure of fecal production is required so those methods also will be considered. Throughout this review attention will be given to estimates of numbers of animals required for sampling quantity and quality of the diet, based primarily on studies with domestic animals. An appendix is provided of the derivation and interrelation of equations of interest.

**Definitions.** Numerous specialized terms are used in this review; for clarity some of these terms are defined as follows:

*Forage*—the vegetation actually grazed by or fed to an animal.

*Herbage*—the vegetation available to a grazing animal (herbage is more inclusive than forage).

*In vitro microdigestion*—refers to the artificial rumen techniques for estimating forage digestibility.

*In vivo microdigestion*—refers to the nylon, silk or dacron bag method for estimating forage digestibility.

*Macrodigestion*—digestion estimates obtained from the entire animal, i.e., by means of indicator techniques and total collection procedures or by standard digestion trial procedures.

*Microdigestion*—an inclusive term referring to both *in vivo* nylon bag and *in vitro* artificial rumen methods of estimating digestibility; represents digestion estimates based on small samples and only part of the animal's digestive tract.

**Methods for studying diet.** Four society monographs appearing in recent years review methods of studying quality and quantity of the diet of grazing animals (Sell *et al.*, 1959; American Society of Animal Production, 1959; Joint Committee, 1962; and Agricultural Board, 1962). About seven general methods are available for estimating botanical or chemical composition of the diet of grazing animals; as follows:

(1) observing free-grazing or tethered (Mitsumata *et al.*, 1959) animals to note the relative abundance of different plants in their diets,

(2) estimating in plots the production and utilization of different species to calculate the diet grazed,

(3) clipping plots before and after grazing to determine use by the difference,

(4) plucking plant units before and after grazing to determine chemical composition and botanical composition by difference,

(5) using esophageal and ruminal fistulated animals to collect samples of the forage grazed, then analyzing these samples,

(6) killing animals to analyze rumen or intestinal tract contents for botanical composition (often used with wild animals), and

(7) examining feces microscopically for botanical residues to indirectly determine dietary intake.

Contrasting and non-supported statements concerning dietary intake are found through the literature because different results have been obtained with the various techniques. Paradoxically, many of these statements are based on observational data recorded to four or five 'significant' figures.

It is generally acknowledged that hand sampling techniques, observational techniques, and stomach analyses are inadequate methods for accurately evaluating the herbage grazed. The most recent techniques for evaluating diets are esophageal and ruminal fistula sampling. The esophageal fistula has considerable advantage over the ruminal fistula for range sampling (Van Dyne and Torell, 1963). The esophageal fistula appears to be the most valid method thus far developed.

Paunch or stomach analysis methods are used routinely in investigations with many large wild animals. Until more precise methods become available, stomach analysis methods will continue to be used; but attention should be given to the limitations of the technique. Procedures for stomach analyses are outlined in detail by Martin and Korschgen (1963) but essentially involve killing an animal, opening the rumen, taking a sample, and analyzing the sample under a microscope. The botanical composition of the sample is visually estimated (e.g., Chippendale, 1962), read with a point frame apparatus (Chamrad and Box, 1964), or laboriously separated manually (e.g., Talbot and Talbot, 1962). In some instances 'volumetric percentages' are reported (e.g., Morris and Schwartz, 1957), and more recently microscopic analysis techniques include conversion of per cent points to per cent weight (Harker *et al.*, 1964; Heady and Van Dyne, 1965; Van Dyne and Heady, 1965) or conversion of frequency values to per cent weight (Sparks and Malechek, 1967). In some instances the forage samples are simply rinsed and then read microscopically (e.g. Van Dyne and Heady, 1965), but in other methods the sample is dried, ground and plant particles counted using a high-power microscope (Sparks and Malechek, 1967). No thorough study has been made comparing the precision, accuracy, and efficiency of the various techniques.

A basic difficulty with the stomach analysis method lies in the sampling regime. Typically only one or two animals are collected per month, and these animals may be collected at different times of day and may represent different age and sex classes (e.g., Klein, 1962; Anderson *et al.*, 1965; Morris and Schwartz, 1957). Practical considerations limit the number of animals sampled, but, as will be shown below, sampling variability is great and many animals may be needed to estimate individual plant species or groups of plants in the diet with necessary precision.

Especially when plants are succulent, and especially when the diet is composed of woody and herbaceous plants, the differential rate of digestion of different plant species can lead to highly variable results when stomach samples are collected at varying times after the plants were grazed. These phenomena may be partially corrected by sorting 'fresh' materials from 'old' materials in the rumen or stomach sample as was done by Talbot and Talbot (1962). Even so, there is considerable subjectivity and tedious labor in differentiating fresh *vs.* old materials. Because of differential rates of digestion of different plants, because of the uncertainty of when the samples were grazed, and because of the modifying and contaminating influences of saliva and bacteria in the samples, chemical or nutritive analyses of forages taken from stomach samples may lead to erroneous conclusions (Norris 1943, and Bissell, 1959). Even considering such limitations, chemical analysis, of the rumen samples has proven useful in wildlife studies where there were large differences in diets of animals from different areas (Klein, 1962).

The most accurate methods for sampling the diet of grazing animals involve the use of esophageal or ruminal fistulas. The esophageal fistula has been used in a wide variety of domestic animals (Van Dyne and Torrell, 1964) and has potential for use in tamed wild animals. Techniques of fistulation and types of cannulae have been continuously improved and estimates have been obtained on the required number of animals needed for sampling chemical and botanical dietary components. For cattle and sheep grazing on dry annual range in early (I, 1490 lb/acre), middle (II, 1220 lb/acre) and late summer (III, 420 lb/acre) with varying amounts of available herbage numbers of animals required for sampling were (Van Dyne and Heady, 1965, *a* and *b*):

| | Sheep | | | Cattle | | | | Sheep | | | Cattle | | |
|---|---|---|---|---|---|---|---|---|---|---|---|---|---|
| | I | II | III | I | II | III | | I | II | III | I | II | III |
| | within 10% | | | with 95% confidence | | | | within 10% | | | with 90% confidence | | |
| Crude protein | 5 | 2 | 1 | 6 | 4 | 3 | Grasses | 30 | 4 | 1 | 10 | 4 | 4 |
| Ether extract | 8 | 1 | 1 | 3 | 3 | 3 | Forbs | 57 | 8 | 6 | 40 | 9 | 7 |
| Lignin | 5 | 1 | 1 | 2 | 1 | 1 | Stems | 7 | 1 | 1 | 1 | 1 | 1 |
| Cellulose | 2 | 1 | 1 | 1 | 1 | 1 | Leaves | 50 | 9 | 4 | 19 | 5 | 4 |

These numbers required for sampling forage with cattle and sheep on the range perhaps approximate those that would be attained with wild animals. It appears fewer animals are required for sampling chemical constituents in the diet than for botanical components. Also, it appears that with a large amount of herbage available more animals are required than when herbage is less abundant. These numbers were obtained using animals of equivalent sex and physiological age, but many factors affect forage selectivity.

Numerous methods of sampling the standing crop of herbage before and after grazing in a given area or of grazed and ungrazed plants have been developed. These methods are not discussed here, for in many instances on rangelands the variability of standing crops and of animal utilization of these herbages is so great that the sampling intensity required for adequate precision is so great as to be impractical. Furthermore, it is difficult to accurately simulate the diet of the grazing animal by observation and hand-plucked samples for chemical analysis. It has been shown repeatedly that the diet grazed usually is of higher quality than the diet selected by man to simulate that of the animal (Van Dyne and Torell, 1964). Methods of botanical analysis of pastures and rangelands are discussed in detail by Brown (1954), Lynch (1960), Forest Service (1959), Phillips (1959), Joint Committee (1962), Agricultural Board (1962), and Grassland Research Institute (1961).

The clearest principle arising from a literature review is 'Forage selectivity is a complex phenomena.' Some of the factors which can influence forage selectivity by the grazing animal include (1) species of animal, (2) age of animal, (3) stage in sexual cycle, (4) competition among animals or stocking rate, (5) intensity of use, (6) level of herbage availability (independent of intensity of use), (7) stage of growth of plants, (8) climatic influences such as wind and temperature, (9) feed supplements, and (10) human influences such as herding or riding. Undoubtedly many of these factors are highly interrelated and many interact. For further discussion of palatability of herbage and animal preference the reader is referred to Heady (1964).

**Determining digestibility in the field, barn, and laboratory.** In investigations with both wild and domestic animals, there are opportunities for studies in the field, the metabolism barn, and in the laboratory. In barn or corral trials wild animals often are used to determine preferences for forages (Webb, 1959; Nichol, 1938; Alkon, 1961; Smith, 1950b and 1953); or the digestibility of the forage is measured by feeding the animal and then collecting fecal samples (Smith, 1950a, 1952, 1957; Smith et al., 1956; Bissell et al., 1955; Bissell and Weir, 1957; French et al., 1955; Magruder et al., 1957; Dietz et al., 1962; Ullrey et al., 1964; and Foot and Romberg, 1965). In many instances it may be possible to combine wild and domestic animals in nutritional studies if the comparative grazing habits and digestive efficiencies are known, so that the techniques of digestion trials and animal variation in the dry lot will be reviewed. Since there have been no digestion trials with grazing wild animals, the information on grazing trials pertains only to domestic animals. Because micromethods for evaluating digestibility of forages have been receiving increasing attention and because these techniques have promise for studies with large wild animals, these methods will be considered in some detail.

**Review of digestion trial methods and animal variation.** Animal nutrition textbooks and methodology monographs have not given much consideration to major sources of variability in digestion trials (Maynard and Loosli, 1962; Crampton and Lloyd, 1959; Kleiber, 1961; American Society of Animal Production, 1959). The two recent monographs on range and pasture research do not discuss either the length of digestion periods required or the number of animals required for various degrees of accuracy (Agricultural Board, 1962; Joint Committee, 1962). General comments in the above books indicate digestion trials should have from 7 to 14 days of collection, and the preliminary period should vary inversely with the degree of change in diets. Crampton and Lloyd (1959) give estimates of variation in dry matter digestibility within humans, rats, and swine, but do not discuss ruminant variability.

Major variations in estimates of digestibility arise from errors in estimating amount and composition of intake and excretion. Variation in estimates of composition of intake under range grazing has been presented in the previous section. This variation is an important part of between-trial differences on the same feed or forage. Within trials several other sources of error may occur as follows: (1) length of preliminary and collection period, (2) between- and within-class variation in digestive ability, (3) level of feeding and nutrient balances, and (4) sampling and analytical error. These factors are discussed in the following sections.

**Preliminary and collection period.** The animal is adjusted to wearing a fecal collection harness, and residues of feed not being investigated are cleared from the digestive tract during the preliminary periods. Because food passage rates are difficult to determine in ruminates, digestion trials usually are made under the assumption of steady-state conditions. Thus, a preliminary period also is required to develop a steady state condition.

Forbes and Grindley (1923) stated there was a lack of good data on the required lengths of preliminary and collection periods. They suggested on the basis of judgement and experience, preliminary periods should be greater than 10 to 20 days. The required length of the preliminary and collection period for domestic ruminants has been investigated by various workers:

**Length of preliminary and collection periods**

| Reference | General results |
| --- | --- |
| *Preliminary period:* | |
| Kennedy and Dinsmore (1909) | hand collected range plants fed to sheep often required more than 3 days before accepted |
| Lloyd *et al.* (1956) | ether extract digestion estimation requires longest preliminary |
| Nicholson *et al.* (1956) | 7-day preliminary adequate if no major ration change |
| Stielau (1960) | least time needed respectively for crude protein, dry matter, crude fiber, and ether extract |
| | |
| *Collection period:* | |
| Staples and Dinusson (1951) | 7 and 10 days comparable except for nitrogen-free extract |
| Reid *et al.* (1952) | 4-, 7-, and 10-day collections comparable for pasture trials |
| King *et al.* (1960) | 6- and 10-day collections comparable |
| Clanton (1961) | 7 and 10 days comparable for digestion coefficients, metabolizable energy, and nitrogen retention |
| Grainger *et al.* (1960) | 3 days give lower dry matter digestion coefficients than 7 days |

In range digestion trials preliminary periods are less critical than collection periods because the test animals normally remain on the range most of the year. In most seasons the dietary changes within a trial are gradual. The length of the collection period is important, especially in minimizing end-point error. End-point error will decrease in inverse proportion to the square root of the number of collection days (Blaxter *et al.*, 1956).

There are few data on the effect on animals of bagging for fecal collections. Hill *et al.* (1961) found over-stimulated fecal production in heifers bagged for short durations, but it is logical that this effect was only temporary. In many grazing and standard digestion trials, animals were bagged for long durations. Topps (1962) bagged 3 sheep for 35 days, Axelsson and Kivimäe (1951) made collections from wethers in crates for 44 days, Greenhalgh *et al.* (1960) bagged steers for at least 50 days and indicated they showed no discomfort, Raymond *et al.* (1953) bagged sheep 150 days with no bad effects, and Vercoe *et al.* (1961) kept three sheep in digestion cages for 52 continuous weeks.

Few range digestion trials have been conducted. Cook and co-workers have conducted sheep digestion trials on pure stands and mixed range plant types in Utah. Their procedures have varied considerably over the last 15 years. Cook and Harris (1950) conducted trials wherein wethers, lambs, and yearlings were grazed with a band, and fecal collections were taken for about 21 days after a 4-day preliminary period. Later, they used wethers (three to four years old) in 8-day preliminary and 6-day collection periods (Cook and Harris, 1951; Cook *et al.*, 1954; and Cook *et al.*, 1956). Recently they shortened their trials to 6-day preliminary and 5-day collection periods (Cook *et al.*, 1961; Cook *et al.*, 1962). In some studies they did not bag sheep during preliminary periods (Cook *et al.*, 1961).

*Within-class variation.* Most workers acknowledge animal variability in digestive efficiency, but the magnitude of this variability and how it changes under different conditions has not been thoroughly evaluated. Animal numbers in standard digestion trials vary from 3 to 20 or more. Forbes *et al.* (1946) fed 22 sheep for 10 days in crates on a 13% crude protein clover-timothy hay and measured the variability of feed intake and digestibility. Since their feed intake variation was relatively low, the difference in digestibility represents variations in fecal composition. Coefficients of variation (among sheep) calculated from their data are as follows: crude protein, 3%; ether extract, 4%; crude fiber, 3%; nitrogen-free extract, 2%; energy, 4%; soluble carbohydrates, 4%; and TDN, 2%. Lignin averaged $-1$% digestibility but had about the same standard deviation as ether extract.

Bartlett (1904) found that, although sheep were smaller and easier to handle in digestion trials, they were nervous, restless, and liable to lose their appetite when confined to stalls; and in parallel experiments with two or more animals, sheep were subject to wider variation than cattle. Forbes and Grinkley (1923) indicated that not less than three and, if practicable, five or more animals should be used to test a feed in digestion trials to account for variability.

Schneider and Ellenberger (1927) found fecal excretion rates of dairy cows were more variable under maintenance than under high feed intake. This agrees with Blaxter *et al.* (1956) who indicate the error attached to digestion coefficients drops with increasing food intake. Jordan and Staples (1951) and Alexander *et al.* (1962) found digestive variation among sheep on constant intakes was greater than among cattle.

Schneider and Lucas (1950) found, in a literature review, that 10 to 25% of the investigated variance components was due to variations in trials. Their between-trial component was due mainly to animal variations, but also to chemical and minor sampling errors. Their data show sheep are slightly more variable than cattle, horses are more variable than sheep, and swine are less variable than cattle. They concluded only few animals need to be used in evaluating a feed, but they do not recommend specific numbers. Groenwald *et al.* (1950) also found animal variation in digestion trials was small and found as much difference in periods of digestion of the same feed as among animals on the feed. Grassland Research Institute workers (1961) have standardized trials to use only three sheep per feed under highly controlled feeding conditions. Schneider *et al.* (1955) used only three sheep per treatment in studying digestibility of pasture.

*Between-class variation.* Various workers have reported forage digestion differences within and between cattle and sheep, but these differences may appear for only certain constituents and feeding regimes. This is reflected by the statements of Cipolloni *et al.* (1951) who said that, '. . . it does not seem possible to state outright that sheep have poorer or better digestive powers than cattle . . .' and '. . . differences between these two species exist and the direction and magnitude of these differences may be functions of the feed and of the nutrient involved.' Taking data from Schneider's book, *Feeds of the World* (1947), for feeds which were reported by two or more authors with nine or more trials per species, Cipolloni *et al.* (1951) compared digestive ability of cattle and sheep. Using covariance to adjust for differences in proximate components, they found that cattle had higher dry roughage digestion coefficients than sheep for organic matter, crude fiber, nitrogen-free extract, and TDN. There were significant species $x$ feed interactions in crude protein digestion. Fraps (1912) states animal variation is less than within-feed variation, and Schneider and Lucas (1950) indicate that worker and batch variations in digestion coefficients are greater than animal variations. Thus, to compare digestive variation the level of feeding, type of ration, and type of animal must be considered.

A comparison of cattle and sheep digestion of roughage diets is as follows:

**Roughage digestive ability of cattle and sheep**

| Reference | General results |
| --- | --- |
| Bartlett (1904) | steers $>$ sheep for coarse fodders, but as high within-as between-class variability |
| Forbes *et al.* (1937) | sheep $>$ cattle, except for crude fiber (CF) (producing cows vs. sheep at maintenance) |
| Watson *et al.* (1948) | no average difference over all feeds |
| Jordan and Staples (1951) | steers slightly higher than lambs on prairie hay |
| French (1956) | East African hair sheep=Zebu oxen, sub-maintenance CF digestion super maintenance |
| Alexander *et al.* (1962) | sheep $>$ cattle on low quality grass hay |
| Dijkstra *et al.* (1962) | wethers=dairy cows |

In most instances cattle digest the fibrous components of low quality roughages somewhat better than sheep. However, there are often as high within-class variations as between classes. Thus, Weston (1959) found strong-wool Merinos digest nitrogen better and crude fiber poorer than fine-wool Merinos. Howes, *et al.* (1963) found Brahmans had higher crude protein digestion coefficients than did herefords when fed low quality diets.

*Application to range conditions.* The studies under dry-lot feeding indicate considerable variation may be found within and between classes of stock. Under range or pasture grazing, in addition to inherent differences in digestive capacity, there are also differences in amount and quality of the feed consumed. A second major difference between range and dry-lot digestion trials is that separate groups of animals often are used in range trials to estimate diet composition and fecal excretion rate and composition. For example, estimates of dietary composition may be secured from a group of esophageal fistulated animals, whereas estimates of excretion rate and composition may be secured from a separate group of animals.

Dry matter or nutrient intake is calculated as follows:

$$\text{Digestibility of nutrient} = 1 \cdot 0 - \frac{\% \text{ indicator in diet}}{\% \text{ indicator in feces}} \times \frac{\% \text{ nutrient in feces}}{\% \text{ nutrient in diet}} \qquad (1)$$

$$\text{Nutrient intake} = \frac{\text{fecal weight}}{\text{indigestibility of nutrient}} \tag{2}$$

From the above equations (derived in Appendix A) it is clear that variations in estimates of digestibility and intake could be due to differences in (1) dietary composition, (2) excretion rate, and (3) fecal composition. Indicators may be indigestible components of the feed or introduced materials.

If constant intake and output composition is assumed, then excretion rate variation is the sole determinant of the error in a digestion coefficient (Blaxter *et al.*, 1956). It follows that two main types of error occur—an independent error and an end point error. The independent error—variations in digestive efficiency during a trial—also includes analytical and sampling errors. End point errors are caused by the irregularity in excretion rates. Both independent and end point errors are inversely proportional to the length of the collection period, but the latter will decrease more rapidly than the former as the period is lengthened (Blaxter *et al.*, 1956). If intake is constant, errors in determination of digestion coefficients are independent of the magnitude of the co-efficients.

Animal differences in fecal excretion and composition are important in range digestion trials for three reasons:

(1) Generally there has been no significant difference in dietary chemical composition from day to day within a period (Van Dyne and Heady, 1965*b*), and probably digestibility of intake is relatively uniform. Thus variations in excretion rate and composition become especially important in assessing estimates of intake and digestion under range conditions.

(2) Because of the constantly changing environment, experimental designs which account for individual animal differences (e.g., latin square and cross-over) are much less useful under range than under dry-lot conditions. Therefore, a knowledge of the magnitude and changes of individual animal differences under varying grazing conditions are especially important.

(3) For the same number of animals, range digestion trials are more expensive than those conducted under dry-lot conditions. Therefore, it is important to assess individual animal variability in order to calculate the minimum number of animals required for any given level of accuracy.

Estimates of variability in fecal excretion rate and composition are calculated here from unpublished data in the experiments of Van Dyne and Lofgreen (1964) and Van Dyne and Meyer (1964*b*). In both range and feedlot trials more sheep than cattle would be required to sample excretion rate.

Usually five steers would estimate excretion rate within 10% of the mean with 95% confidence. Sheep variability was greatest with a high availability of herbage, but cattle variability was greatest with low availability of herbage. These estimates are based on a seven-day preliminary period and a seven-day collection period. Fecal ether extract, other carbohydrates, and silica-free ash generally would require from four to nine animals. Crude protein, cellulose, and lignin in the feces would require a maximum of two to five animals to sample within 10% of the mean with 95% confidence.

In lieu of total collection of feces from wild animals, the common practice is to determine the number of pellet groups in an area for a given period of time and convert these data to animal days of range usage (Leopold *et al.*, 1951; Nichol, 1938; Julander *et al.*, 1963). There are many limitations to such techniques. The level of forage intake and the kind of forage eaten both influence the defecation rates for mule deer (Smith, 1964) and probably for other wild animals. Coarse woody plants produce higher defecation rates than the same amount of more succulent and more digestible herbaceous plants. There are also many difficulties in making accurate and precise counts of pellet groups in the field. Even a few heavy rains can cause movement or dislocation of pellet groups from study plots (Wallmo *et al.*, 1962). On an area with a high deer population, Smith (1964) found on a winter range that more than 200 plots of 100 ft.$^2$ each may be required for reasonable precision.

*Review of methods for digestion trials with grazing animals.* Digestibility and intake of forage are determined by indirect methods in range and pasture studies. These methods are variously called indicator techniques, fecal index techniques, ratio techniques, etc., according to the special type of usage. General reviews of methodology are given by Schneider *et al.* (1955), Reid and Kennedy (1956), Corbett (1960), Brisson (1960), and Reid (1962). Special consideration has been given ratio and tracer methods by Raymond (1954), lignin by Salo (1957a, 1957b, 1958) and Milford (1957), and to fecal index methods by Raymond *et al.* (1954), Greenhalgh and Corbett (1960), and Minson and Kemp (1961). Using either an internal indicator (e.g., lignin; Garrigus, 1934; Forbes and Garrigus, 1948) or using the forage: feces ratio of nitrogen concentration can have disadvantages. Lignin, silica, and fecal nitrogen as indicators are reviewed with special consideration given to their use in range investigations.

*Lignin ratio technique.* Disadvantages of the lignin ratio procedure according to Milford (1957) are: (1) lignin is not a distinct chemical entity, (2)

TABLE 5.1. Examples of range digestion trials in the United States using lignin ratio technique

| Reference | State | Type of range | Season* | Class** of Stock | Sample collection procedure | |
|---|---|---|---|---|---|---|
| | | | | | Forage | Feces |
| Burns (1958) | Idaho | crested wheatgrass | Sp–F | C | hand clipping | total collection |
| Cook and Harris (1950) | Utah | salt desert shrub | W | S | hand plucking | total collection |
| Cook and Harris (1951) | Utah | salt desert shrub | W | S | hand plucking | total collection |
| Cook and Stoddart | Utah | wheatgrass pastures | Sp | C & S | hand plucked | sample from pate |
| Cook et al. (1954) | Utah | salt desert shrub | W | S | hand plucking | total collection |
| Cook et al. (1956) | Utah | wheatgrass pastures | Sp–Su | S | hand plucking | total collection |
| Cook et al. (1961) | Utah | mountain range | Su | S | esophageal fistula | total collection |
| Cook et al. (1962) | Utah | salt desert shrub | W | S | esophageal fistula | total collection |
| Cook et al. (1963) | Utah | mountain range | Su | C & S | esophageal fistula | sample from pate |
| Hale et al. (1962) | Ga. | wiregrass-pine | Sp–F | C | hand clipping | grab samples |
| Halls et al. (1957) | Ga. | wiregrass-pine | Year | C | hand clipping | grab samples |
| Smart et al. (1960) | N. Car. | Arundinaria spp. | Year | C | hand clipping | grab samples |
| Van Dyne (1960) | Mont. | foothill bunchgrass | W | S | esophageal fistula | total collection |
| Van Dyne (1960) | Mont. | foothill bunchgrass | W | C | esophageal fistula | total collection |
| Wheeler (1962) | Ore. | crested wheatgrass | Sp–F | C | hand clipped | total and grab spl. |
| Van Dyne and Lofgreen (1964) | Calif. | annual range | Su | C & S | esophageal fistula | total collections |

* Sp, Su, F, W, and year = refer to the four seasons and yearlong.
** C = cattle; S = sheep.

impurities may become attached to lignin during chemical analysis, (3) methods of lignin analysis are tedious and expensive, (4) selective grazing can introduce high errors in sampling of forage actually consumed, (5) lignin may be partially digestible, and (6) changes in chemical composition of lignin may occur through the digestive tract. Albeit there are many imperfections in the lignin ratio procedure, most United States range investigators have used the lignin ratio technique to calculate forage indigestibility or total forage intake. These trials are summarized in Table 5.1. Chromogen was used in addition to lignin but found unsatisfactory by Cook and Harris (1951) and Van Dyne (1960c). Smart *et al.* (1960) found lignin and silica always gave lower TDN values than copper-chlorophyll pigments as indicators.

The lignin ratio procedure has given reasonable estimates of forage intake and digestion in most range studies when a good estimate was available for lignin composition of forage grazed. But because it is difficult to harvest range herbage and feed it to animals in dry-lot digestion trials, the possibility of lignin digestibility has not been investigated for most range herbages. There is some indication in the studies reviewed that lignin may be partially digested in immature herbages.

*Silica ratio technique.* Silica has been used in a number of investigations as an inert indicator, and these studies are summarized as follows:

**Use of silica as an inert indicator in roughage trials**

| Reference | Year | Species | General value or use |
| --- | --- | --- | --- |
| Druce and Wilcox | 1949 | rabbit | highly variable recovery |
| Forbes and Beegle | 1916 | cattle | incomplete recovery |
| Gallup and Kuhlman | 1931 | cattle | compared well with conventional |
| Gallup *et al.* | 1945 | steers | more than 100% recovery |
| Goncarova *et al.* | 1960 | unspec. | used in pasture study with $Fe_2O_3$ |
| Hodgson and Knott | 1932 | cattle | gave reasonable results |
| Knott *et al.* | 1936 | cattle | too much soil contamination (dust) |
| Platikanoff and Popoff | 1937 | sheep | compared well with conventional |
| Skulmowski *et al.* | 1943 | unspec. | compared well with conventional |
| Smart *et al.* | 1960 | cattle | compared well with lignin |
| Wildt | 1874 | sheep | 8—14% absorbed |
| Van Dyne & Lofgreen | 1964 | cattle, sheep | unsuitable for annual range |

In general, the silica ratio has not been successful for grazing studies because, in the past, it has been impossible to sample the same forage as does the animal. Because of the high silica content in soil, even small amount of soil contamination of herbage or fecal samples give variable and invalid results in digestion trials.

There has been some indication that silica is absorbed (Wildt, 1874), but there is no known bodily need for silica (Crampton and Lloyd, 1959). Silica may be absorbed, but it is readily excreted in the feces, probably through the bile, because very little is found in the urine.

*Fecal nitrogen index.* The basis of the nitrogen index ratio procedure, originated by Lancaster (1949), is that total fecal nitrogen is directly proportional to the nitrogen content of the diet. Fecal nitrogen is composed of undigested forage nitrogen and of metabolic nitrogen. Metabolic nitrogen is excreted in amounts approximately proportional to the dry matter intake of the animal (Blaxter and Mitchell, 1948; Hutchinson, 1958). Fecal chromogen and crude fiber have also been used in these fecal index ratios.

The main advantage of the fecal index procedure is that an estimate of the diet of the grazing animal is not required. But it does require digestion trials to develop the relation between forage and fecal nitrogen for the forage under study. The obvious difficulty with this procedure under most range conditions is the impracticability of obtaining enough herbage with which to conduct the digestibility trials. It is also difficult to get animals to consume low quality roughages in standard digestion trials. The fecal index techniques are also subject to inaccuracies due to inadequate fecal sampling when total collections are not taken.

The sources of error in fecal index regression techniques have been summarized by Greenhalgh and Corbett (1960): (1) different herbages may have different nitrogen digestibilities for the same organic matter digestibility, (2) differences between animals exist in digestibility, and (3) errors of measurement are important. Generally the first type of error—differences between herbages—is greater than differences between animals or errors of measurement.

The use of the nitrogen fecal index technique in grazing studies, where the animals did not receive supplements, is summarized in Table 5.2. Fecal nitrogen has been used to predict both the feed:feces ratio and the organic matter digestibility. For purposes of comparison, the similarities of these equations are shown as follows:

$$\text{let } OM = \text{organic matter digestibility}$$
$$a, b = \text{constants of linear regression equations}$$

$E_{om}$ = weight of organic matter in the feces
$F_{om}$ = weight of the organic matter in the feed
N   = nitrogen content of the feces

The usual form of the equations are

$$OM = a + b \times N \tag{3}$$

$$\frac{F_{om}}{E_{om}} = a^1 + b^1 \times N \text{ (Lancaster, 1949)} \tag{4}$$

From equation (3) of Appendix A, it is seen

$$OM = \frac{F_{om} - E_{om}}{F_{om}} \times 100 \tag{5}$$

It is easily seen that organic matter digestibility may be represented as

$$\frac{F_{om} - E_{om}}{F_{om}} \times 100 = a + b \times N \tag{6}$$

By successive algebraic manipulations, it can be shown

$$\frac{F_{om}}{E_{om}} = \frac{1}{1 - \cdot 01a - \cdot 01b \times N} \tag{7}$$

The feed to feces ratio for an assumed 2% nitrogen in the feces is calculated for each equation in Table 5.2. Estimates of feed:feces vary from 2·00 to 3·33. Most of these equations were developed with animals fed palatable herbages of more than 8 to 10% crude protein.

Some of the equations in Table 5.2 were developed using $Cr_2O_3$ and grab sampling to estimate total fecal output, and others are based on total collections. The $Cr_2O_3$ grab sampling procedure can lead to considerable error if the pattern and changes of diurnal excretion rates under grazing conditions are not determined and considered in the sampling scheme (Raymond, 1954; Lambourne, 1957; Van Dyne, 1960; and Wheeler, 1962).

The equations of Minson and Kemp (1961) show even for high quality herbages the within-season influence is important. This is also true for

TABLE 5.2. Examples of the use of fecal nitrogen index technique

| Reference | Country | Forage | Class | Prediction equation* | F/X for 2% N |
|---|---|---|---|---|---|
| Elliot & Fokkema (1961) | Rhodesia | veld grassland | cows | $F/X = ·48 + 1·04N$ | 2·00 |
| Fels et al. (1959) | (from literature review) | | sheep | $F/X = ·64 + 1·02N$ | 2·68 |
| Fels et al. (1959) | Australia | mixed annuals | sheep | $F/X = ·97 + ·86N$ | 2·69 |
| Greenhalgh & Corbett (1960) | Scotland | grasses first growth | steers | $F/X = 1/(·55 − ·11N)$** | 3·03 |
| Greenhalgh & Corbett (1960) | Scotland | grasses aftermath | steers | $F/X = 1/(·61 − ·11N)$ | 2·56 |
| Gupta et al. (1962) | India | Heteropogon and Sorghum | bullocks & buffalo | $F/X = ·87 + 1·24N$ | 3·35 |
| Holder (1962) | Australia | mixed annuals | sheep | $F/X = 1·03 + ·24N + ·19N^2$*** | 2·27 |
| Holmes et al. (1961) | England | ryegrass-clover | sheep | $F/X = 1/(·56 − ·13N)$ (late) | 3·33 |
| Holmes et al. (1961) | England | ryegrass-clover | sheep | $F/X = 1/(·56 − ·11N)$ (early) | 2·94 |
| Lancaster (1954) | New Zealand | improved pasture | cattle | $F/X = 1·02 + ·97N$ | 2·96 |
| Minson & Kemp (1961) | England | April grass sward | sheep | $F/X = 1/(·47 − ·06N)$**** | 2·86 |
| Minson & Kemp (1961) | England | Nov. grass sward | sheep | $F/X = 1/(·73 − ·12N)$ | 2·04 |
| Topps (1962) | Rhodesia | Hyparrhenia veld | cows | $F/X = 1/(·73 − ·16N)$***** (1959) | 2·44 |
| Topps (1962) | Rhodesia | Hyparrhenia veld | cows | $F/X = 1/(·58 − ·06N)$ (1960) | 2·22 |
| Vercoe et al. (1961) | Australia | improved pasture | sheep | $F/X = ·98 + ·86N$ | 2·70 |

*F/X is feed to feces ratio and N=nitrogen, all on an organic matter basis. The equations presented by the authors have been rounded to two decimal places.

**Equations were originally given as OM=11N+44·9 and OM=11N+39·4.

***The equation is originally given with negative linear and quadratic coefficients.

****The equations were originally given as OM=53·3+6·05N and OM=27·4+11·74N.

*****The equations were originally given as OM=16·43N+26·55 and OM=6·51N+42·22.

herbages in a Mediterranean environment (Vercoe and Pearce, 1962). The equation of Topps (1962) shows between-season variation even for the same vegetation is important. He stated that slower than normal growth of the grass in one year caused a less rapid increase of fiber and lignin. Although the herbage was mature at the time of the study, there were carry-over effects from the growing season affecting his regression equations for predicting the feed:feces ratios.

The equations of Greenhalgh and Corbett (1960) show how selective grazing by the animal could affect results. Equations were different for first growth and aftermath grasses. They further indicated faulty results may be obtained if an equation developed with sheep is applied to cattle (as was done by Holmes *et al.*, 1961), or even if an equation developed with beef steers is applied to dairy cows.

**Review of Microdigestion Studies.** Micromethods for nutritive evaluation of forages have received considerable attention in recent years and are reviewed by Annison and Lewis (1959), Barnett and Reid (1961), and Johnson (1963). The application of micromethods to nutritive evaluation of range forages is discussed by Van Dyne (1962). The review herein is concerned primarily with methodological investigations since 1961.

*In vivo microdigestion.* Major types of *in vivo* microdigestion techniques are (1) small bag studies (nylon, dacron, or silk) (Quinn *et al.*, 1938), (2) cotton thread digestion (Hoflund *et al.*, 1948, and Balch and Johnson, 1950), (3) 'vivar' techniques (*in vivo* artificial rumen of Fina *et al.*, 1962). (4) the insertion of prepared foodstuff sticks in perforated capsules placed in the rumen (Protasenija, 1956). The small bag technique has been the most widely used *in vivo* microdigestion procedure. Several variables influencing these estimates have been reported and are discussed in the following paragraphs.

*Fermentation time.* Fermentation times vary from as few as 3 up to 96 hr. Lusk *et al.* (1962) investigated cellulose digestion from 36 to 72 hr. in 12-hr. increments and found digestion increased from 36 to 60 hr. but not thereafter for Bermudagrass hay. A similar trial on alfalfa hay showed no significant difference in microdigestion of cellulose from 36 to 72 hr. Van Dyne (1962) found cellulose digestion had a direct curvilinear relationship with time for fermentation from 0 to 72 hr.; cellulose digestion at 48 hr. was as high as for longer fermentations. Hopson *et al.* (1963) made time studies from

6 to 42 hr. in 6-hr. increments on 1 g samples *in vitro* and *in vivo*. There was a striking similarity between the time curves of these two procedures, and maximum values would have been reached at about 48 hr.

Van Keuren and Heinemann (1962) studied per cent dry matter digestion in a steer grazing alfalfa or orchardgrass pasture or fed alfalfa hay. After 48 hr. fermentation they found little *in vivo* digestibility change in alfalfa or alfalfa-orchardgrass mixtures, but for orchardgrass alone digestion increased up to 96 hr. Archibald *et al.* (1961) used a 48 hr. fermentation for 30 g samples of finely chopped hay (·75 inch) but found microdigestion estimates for carbohydrate constituents were less than those for total collection, probably because the forages in the bags did not benefit from chewing and rumination.

In summary, depending upon the size, grind, and type of sample, about a 48-hr. *in vivo* fermentation period gives results within a few per cent of total collection per cent cellulose digestion estimates.

*Amount and size of sample.* Samples vary from finely ground (40-mesh) to chopped forage (·75 inch). Particle size seems to be relatively unimportant as Van Keuren and Heinemann (1962) found 20-, 40-, and 60-mesh grinds had very little influence on microdigestion estimates. Kercher (1962) compared alfalfa hay, native hay, and wheat straw samples ground through ·25 inch, 1 mm, and 2 mm screens. There was no significant difference in dry matter disappearance due to fineness of grind.

Sample size is usually inversely related to digestion. Van Dyne (1962) found sample size, from 2 to 10 g, was inversely related to per cent cellulose digestion. Van Keuren and Heinemann (1962) found increasing sample size from 5 to 10 g depressed dry matter digestion only for less than 48 hr. fermentation.

*Variability.* The study of Archibald *et al.* (1961) with four ruminal fistulated cows showed low animal variability for *in vivo* microdigestion. Tests made on consecutive days in one steer by Van Keuren and Heinemann (162) show highly uniform results among days. Van Dyne (1962) found significant individual animal variation for microdigestion of cellulose in a mixed annual range forage sample when the animals were fed a low quality oat hay; however, sampling error decreased with increasing fermentation duration. Hopson *et al.* (1963) also found sampling error decreased as fermentation time increased. Filter paper digestion in nylon bags has recently been studied on diets of wheat straw alone or wheat straw with clover fed to buffalo and Zebu cattle (Ichhponani *et al.*, 1962). These authors found differences between animal species were estimated similarly with *in vivo* and *in vitro* microdigestion techniques; also, animal variation was greater on the lower quality diet.

*Physical loss from bags.* Since nylon bags vary from approximately 80 to 120 threads per inch, a considerable amount of material could be lost by physical means. Archibald *et al.* (1961) tested physical losses by placing bags in a water bath with a small stirring motor. About 25% of alfalfa and 20% of timothy samples were water soluble in 48 hr. by this treatment. However, they did not determine if insoluble material passed through the weave. Lusk *et al.* (1962) subjected eleven forage samples in nylon bags to vigorous agitation for 18 hr. in a water bath. This resulted in an average loss of 0·24% of the cellulose. The same procedure with Solka-floc (BW 40) resulted in 15 to 30% losses, but about 90% of BW 40 Solka-floc will pass through a 100-mesh screen.

Post soaking of bags upon removal from the rumen increased per cent dry matter digestion estimates but did not reduce the coefficient of variation in studies by Van Keuren and Heinemann (1962). Bags containing forages were placed in a running water for 24 hr. by Hopson *et al.* (1963) to determine physical loss. Approximately 1% of the forage dry matter was lost in this manner. Van Dyne (1962) found rinsing treatment (post-digestion) increased dry matter digestion estimates more than cellulose digestion estimates. Lusk *et al.* (1962) placed empty nylon bags in the rumen and found 2·9 mg. cellulose entered the bag from the ingesta. But this addition would change cellulose digestion estimates only by 0·1%.

The above studies indicate losses through the weave of the bag are negligible for most forages even when ground through a 40-mesh screen. But purified cellulose sources which are even finer (e.g., Solka-floc) may suffer appreciable physical losses.

*Comparison to total collection.* Some workers (e.g. Lusk *et al.* 1962) have found there were no significant differences between micro and macrodigestion. But comparisons of micro- and macrodigestion are difficult to make because differences in base diets have been very important in modifying microdigestion estimates (Hopson *et al.*, 1963, and Van Dyne, 1962).

Extrapolation of the data of Hopson *et al.* (1963) shows *in vivo* cellulose digestion at 48 hr. would be within 2 to 3% of the total collection coefficients. But Archibald *et al.* (1961) found total collection cellulose digestion estimates were from 1 to 1·3 times higher than dacron bag estimates. However, these workers used 30 g samples of ·75-inch chop forages, and they placed each bag containing samples in a larger bag of the same forage material. Still they found micro- and macrodigestion techniques ranked all constituents in two forages the same except for dry matter and lignin. Microdigestion failed to distinguish between the two forages in dry matter digestion and gave lignin digestion coefficients two to three times higher than macrodigestion procedure. Microdigestion estimates of crude protein digestion were higher than total collection because of the presence of metabolic nitrogen in the feces. Ether extract microdigestion estimates were higher than macrodigestion, probably because insoluble calcium soaps are found in the fecal material. Lignin macrodigestion varied from 0 to 26·4% and by microdigestion technique, it varied from 17·3 to 33·5%.

**In vitro microdigestion.** *In vitro* microdigestion techniques are diverse, but artificial rumen systems may be categorized as: (1) continuous flow, (2) semi-permeable membrane, (3) all-glass volumetric, or (4) all-glass gravametric systems. All-glass gravametric systems are of major interest herein because of their simplicity and usefulness in forage evaluation studies (Van Dyne, 1963).

Numerous sources of variation exist in *in vitro* procedures including: (1) base feed of the animals from which inocula are obtained, (2) type and amount of buffer, (3) relative amounts of rumen liquor, substrate, and energy source in media, (4) supply of vitamins and trace minerals, (5) particle size of samples, (6) length of fermentation, and others. Probably no two laboratories use exactly the same procedures, and it may not be desirable that standardization takes place. Each laboratory must develop its own technique and consider or control the above variables. Many important variables affecting *in vitro* artificial rumen fermentations are discussed in the general references above and in a review by Van Dyne (1967), but some additional aspects are discussed below.

*Microbial variations.* One of the variables affecting *in vitro* fermentations is the concentration of bacteria and protozoa in the inocula. There is evidence of diurnal, daily, weekly, and among-animal variation in ruminal microbe concentration and activity (Williams and Christian, 1956). These workers evaluated differences in ruminal microbial end products, which indirectly reflect variations in microbial numbers. Sampling errors for ruminal microbial numbers are large in range and pasture studies due to the continually changing environment. However, differences between sheep were relatively constant over a period of time. Williams and Christian (1956) show that it would be preferable to take single samples from as many sheep as possible rather than repeat samples on the same animal.

Even under dry-lot feeding it has been shown there are differences in ruminal microbe numbers due to feed treatments, animals, days, and also day × treatment interactions (Purser and Moir, 1959). Other workers have indicated that weekly variations are so great that they may completely mask seaonal changes (Kamstra and Miller, 1960). The fact that diurnal variation in numbers of microbes exists has been well documented (Hungate *et al.*, 1960; Moore *et al.*, 1962). Generally, most uniform microbial conditions are found from two to four hours after feeding.

Bowden and Church (1962*a*) measured within- and between-trial variation in 48-hr. *in vitro* fermentations from a single source of inocula. Their data show within-trial dry matter digestibility variation for most substrates is relatively small; variation in cellulose digestibility was only slightly higher. One factor affecting between-trial variability was the pattern of watering the fistulated steer from which the inocula was obtained. It may be difficult to control the watering of animals grazing on the range and used to provide inocula samples; thus concentrations of bacteria in the ruminal fluid might vary considerably. Therefore, any sampling pattern under range conditions should take into consideration these behavioral aspects. Baumgardt *et al.* (1962*a, b*) maintained a cow on a constant diet and sampled on four different days in a 2-month period; they found a coefficient of variation in digestibility of about 1·7 between dates.

For range investigations it appears that as large a number of animals as possible should be used at one time, even at the expense of investigating within- and between-day variations. The time of day for sampling should be based on the grazing pattern. Thus, sampling a large number of animals about four hours after grazing should give comparable results in a grazing study.

*Procedural variations.* Several important procedural variations include processing of forage samples, mechanics of the *in vitro* system, and inocula source and processing.

Johnson *et al.* (1962*a*) found a slightly higher correlation in undried than dried forages between 12-hr. *in vitro* cellulose and total collection dry matter digestibilities. However, under range conditions when the forage is low in nutritional value, the herbage generally is dry and mature; thus, this variation is not important.

Caballero *et al.* (1962) compared the *in vitro* procedures of Quicke *et al.* (1959) and Tilley and Terry (1963) to cellulose solubility in cupriethylene diamine. They found dry matter and cellulose digestibility obtained by the two microdigestion procedures gave digestion estimates within a few per cent of the respective values determined by total collection. Cupriethylene diamine cellulose solubility did not differentiate Pangolagrass from Bermudagrass hay; whereas, the *in vitro* and total collection procedures did. This indicates different *in vitro* procedures may give similar results as was also found by Church and Peterson (1960), Barnes and Mott (1962), and Van Dyne (1962).

Most pasture workers now study *in vitro* digestibility with inocula obtained from animals maintained on diets similar to the substrates which are being investigated. But in a recent *in vitro* digestibility study Pritchard *et al.* (1963) used goats maintained on alfalfa hay to furnish inocula for *in vitro* digestion of six grass species at eight stages of growth. Variation may be minimized by using a single base feed to provide the inoculum, but results would be more valuable if a forage similar to those digested was fed to the animals used as the inocula source.

The base diet greatly influences the magnitudes of *in vitro* digestibilities (Asplund *et al.*, 1958; Clark and Mott, 1960; Taylor *et al.*, 1960; Van Dyne, 1962). However, inclusion of a standard sample in each fermentation and adjusting all data to the standard may de-emphasize the importance of the base feed (Tilley and Terry, 1963; Van Dyne, 1963).

Within a reasonable range, the amount of substrate and the amount of sample has little effect on *in vitro* cellulose digestibility estimates (Baumgardt *et al.*, 1962*a*).

*Macrodigestion-microdigestion correlations.* Many investigators have correlated artificial rumen or nylon bag digestibilities with those obtained by conventional procedures. But the degree of correlation varied widely among laboratories, forages, and times within the same forage and laboratory. This variation is understandable because the techniques are highly empirical. Several general conclusions can be drawn from these studies: (1) there is a high correlation between micro- and macrodigestion estimates for dry matter and cellulose, (2) the base feed of animals used as a source of inoculum is important, (3) in many instances 48-hr. microdigestion estimates for cellulose are within 2 to 5% of the macrodigestion estimate, (4) the microdigestion estimates are more precise, and (5) better results generally are obtained when the same class of stock and feeds are used in both micro- and macrodigestion comparisons.

*Comparisons under range and drylot conditions.* Nylon bag and artificial rumen techniques were compared to macrodigestion techniques under range and drylot conditions with cattle and sheep by Van Dyne and Weir (1964, 1966). Averaged over all techniques, samples, and periods, about six cattle and four sheep would be required as inocula sources to estimate microdigestion within 10% of the mean with 95% confidence. More animals would be required with a high availability of range herbage than with limited herbage available. Fewer animals would be required for the same precision by microdigestion techniques than by macrodigestion techniques (Van Dyne and Weir, 1964). Range forage samples taken from esophageal fistulated cattle or sheep were digested better when the base feed was pelleted alfalfa than when it was range forage. Digestion of an alfalfa sample was less affected by the base diet. Solka-floc, a purified wood cellulose, was digested better *in vitro* when the inocula came from animals grazing on the range than from animals fed pelleted alfalfa. Solka-floc was of limited value as a standard sample for the *in vivo* trials because seeds apparently ruptured the nylon bags enough to allow passage of the finely divided material through the wall of the bags.

**Methods for determining forage intake.** The determination of the quantity of forage consumed by grazing animals usually is dependent on knowledge of total fecal output. Total fecal output can be measured in domestic animals by equipping them with a harness and collection apparatus. This procedure might be used with tamed wild animals but obviously will have limitations. Another procedure of measuring fecal output is to regularly dose the animal with an indigestible indicator material so that a steady-state of excretion of the indicator is reached. Then the amount of the indicator in a grab sample of the feces can be used to calculate total fecal output as follows:

Total fecal output/day $=$

$$\frac{\text{amount of indicator administered per day}}{\text{amount of indicator in grab sample}} \times \text{weight of grab sample} \quad (8)$$

The most commonly used indicator material is chromic oxide, $Cr_2O_3$. Although many studies have been made with this indicator, and although improvements are being made in getting more uniform excretion of the indicator, few workers have found this method satisfactory for grazing animals. If recovery of the indicator is complete and if the diurnal variability of concentration of the indicator in the feces is known, then the technique still can be used. Little is known, however, about recovery or excretion curves for animals on rangelands although there is a voluminous literature concerning the use of chromic oxide in nutritional studies.

About five indirect methods have been used to determine forage intake of grazing animals from total fecal output and composition, or from body weight changes.

*Ratio techniques.* The lignin and chromogen ratio procedures (Forbes and Garrigus, 1948, and Reid *et al.*, 1952) have been used in range studies with two assumptions—that the indicator is indigestible and that an accurate determination of the indicator concentration may be obtained. Chromogen ratio procedures have been inadequate under range conditions (Cook and Harris, 1951; Van Dyne, 1960). Although there are many disadvantages to the lignin ratio procedure (*see* review in section on macrodigestion estimates), it has remained the most widely used method in United States range investigations.

One shortcoming of the lignin or chromogen ratio techniques is that estimates of forage intake are based on amount and fecal composition of one group of animals and diet composition of another. Thus only one valid estimate of intake is available, and there is no measure of reliability.

*Fecal nitrogen index.* The fecal nitrogen index procedure originated by Lancaster (1949) has been used widely to determine forage intake in grazing studies in New Zealand, Australia, Africa, and Great Britain (*see* example equations in the section on macrodigestion studies). The assumptions in this technique are: (1) the herbage cut and fed to animals is similar in composition to that selected by the grazing animal; and (2) the pen-fed and grazing animals digest the pasture material to the same extent. The impossibility of harvesting forage similar to that grazed on the range precludes its use in many arid and rough topography ranges and for many wildlife studies.

*Nitrogen balance.* Pasture intake by grazing sheep was calculated from nitrogen balance by Beeston and Hogan (1960). These workers reasoned that a mature wether, whose weight was not varying appreciably, stored nitrogen only in the wool; therefore, nitrogen intake was equal to the amount in the urine, feces and wool.

To measure intake this way would require: (1) long-term studies to overcome the appreciable variations in excretion; and (2) an assumption about the amount of nitrogen stored in the wool. The difficulty of making long-term feces and urine collections on the range and variations in nitrogen storage in the wool preclude extensive use of this system of calculating intake under western range conditions.

*Metabolic fraction of feces.* Dry matter intake by grazing animals can be calculated from metabolic fecal nitrogen according to Owen (1961). He found a high correlation between dry matter intake and the fecal fraction which dissolves in 0·2N HCl in 18 hrs.

This procedure has not been tested thoroughly and is subject to many of the disadvantages inherent in the fecal nitrogen index techniques.

*Weight balance.* Allden (1962) used animal weight balance in a one-hour period of grazing to estimate herbage intake of sheep. The sheep were harnessed for fecal and urine collection, and insensible weight losses were estimated from harnessed sheep which were not permitted to graze. Such short-term measurements are not applicable to range conditions where an animal's grazing activity varies widely during the day.

*General consideration of methods.* In addition to being accurate and precise, an 'ideal' method for determining forage intake of grazing animals should: (1) be applicable to individual animals rather than groups, (2) be based on measurement of dietary or fecal components which can be easily and accurately analysed, (3) not depend on harvesting range herbages for dry-lot digestion trials, (4) be applicable to both cattle and sheep, and (5) be useable on all types of ranges in all seasons.

None of the above methods meet all these requirements. A new method (Van Dyne and Meyer, 1964a) of determining forage intake of grazing animals will be discussed in this section.

In this method a standard sample of known digestion, alfalfa in this case, is included in each trial. Microdigestion estimates for range forages is then expressed in terms of the standard as follows:

'Adjustment ratio' =

$$\left[ \frac{\text{Microdigestion of range forage}}{\text{Microdigestion of standard forage}} \right] \quad \begin{array}{l} \text{when inocula were from} \\ \text{grazing animals} \end{array} \quad (9)$$

This ratio is then multiplied by the microdigestion value for the standard sample when it was digested by inocula from animals fed the standard forage on dry-lot, as follows:

'Adjusted microdigestion estimate' =

$$\left[ \begin{array}{c} \text{Adjustment} \\ \text{ratio} \end{array} \right] \times \left[ \begin{array}{c} \text{Microdigestion of} \\ \text{standard} \end{array} \right] \quad \begin{array}{l} \text{when inocula were} \\ \text{from animals fed} \quad (10) \\ \text{standard forage} \end{array}$$

The third step is to develop a simple linear regression equation between macrodigestion of the standard forage by total collection procedures (Y) and microdigestion of the standard forage when inocula were from animals fed the standard forage (X) as follows:

Macrodigestion of standard =

$$a + b \left[ \begin{array}{c} \text{Microdigestion} \\ \text{of standard} \end{array} \right] \quad \begin{array}{l} \text{when inocula were from animals} \\ \text{fed standard forage} \end{array} \quad (11)$$

The next step is to use the 'adjusted microdigestion estimate' as the X value and calculate 'predicted macrodigestion estimate,' as follows:

'Predicted macrodigestion estimate' =

$$a + b \left[ \begin{array}{c} \text{'Adjusted} \\ \text{microdigestion} \\ \text{estimate'} \end{array} \right] \quad (12)$$

In practice, these steps are combined into one equation as follows:

'Predicted macrodigestion estimate' =

$$a + b \left[ \frac{\text{Microdigestion of range forage}}{\text{Microdigestion of standard forage}} \right] \times \left[ \text{Microdigestion of standard forage} \right] \quad (13)$$

when inocula were from grazing animals      when inocula were from animals fed standard forage

These predicted macrodigestion estimates were then used to calculate feed intake by using the equation 24 of Appendix A.

This procedure holds special promise for calculating individual animal forage intake. Estimates of forage composition, fecal output and composition, and microdigestion all could be obtained from a bifistulated animal. Thus, forage intake estimates could be made on an individual rather than a group basis as described herein. A group basis was used herein because in calculating forage intake part of the estimates usually come from esophageal fistulated animals and part from the ruminal fistulated animals.

## Appendix A

### Equations for Calculating Digestion and Intake of Range Forages

For purposes of discussion, the following symbols are defined:

A  = Amount external indicator administered per day.
$A_E$ = Concentration of external indicator in a sample of feces.
$C_i$ = Concentration in per cent of a digestible nutrient in the diet.
$D_i$ = apparent digestibility coefficient in per cent of a nutrient.
    $D_{dm}$ = Digestibility of dry matter.
    $D_c$  = Digestibility of cellulose.
E   = Fecal excretion weight per day.

$E_s$    = Weight of a sample of feces.

F  = Forage intake weight per day.

I   = Indicator concentration in per cent.

$I_E$    = Indicator in feces.

$I_F$    = Indicator in forage.

$M_i$ = Microdigestion estimates for nutrient *i* in per cent.

$N_i$ = Nutrient concentration in per cent.

$N_{iE}$  = Nutrient in feces—then $N_{cE}$ = fecal cellulose.

$N_{iF}$  = Nutrient in forage—then $N_{cF}$ = forage cellulose.

If an indicator is truly indigestible, it follows that the amount consumed is equal to that excreted, or

(1)   $F \times I_F = E \times I_E$

Dividing by $I_F$ leads to

(2)   $F = E \times \dfrac{I_E}{I_F}$

By definition

(3)   $D_{dm} - \dfrac{F - E}{F} \times 100$

by arrangement

(4)   $D_{dm} = 100 - \dfrac{E}{F} \times 100$

But from equation (2) F may be substituted, therefore

(5)   $D_{dm} = 100 - \dfrac{E}{E \times I_E} \times 100$
$\phantom{(5)   D_{dm} = 100 - \dfrac{E}{}}{}\overline{I_F}$

This simplifies to

(6)   $D_{dm} = 100 - \dfrac{I_F}{I_E} \times 100$

Now if $D_{dm}$ is known, then $(1 - \dfrac{D_{dm}}{100})$ is the indigestibility of a forage. Then the relation between F and E is

(7)   $E = F \times (1 - \dfrac{D_{dm}}{100})$

which rearranges to

(8)   $F = \dfrac{E}{1 - \dfrac{D_{dm}}{100}}$

or to

(9)   $F = \dfrac{E}{100 - D_{dm}} \times 100$

By similar reasoning the digestibility of a nutrient is shown to be

(10)   $D_i = \dfrac{\dfrac{F \times N_{iF}}{100} - \dfrac{E \times N_{iE}}{100}}{\dfrac{F \times N_{iF}}{100}} \times 100$

which can be rearranged to

(11)   $D_i = \dfrac{F \times N_{iF} - E \times N_{iE}}{F \times N_{iF}} \times 100$

which rearranges to a form equivalent to (4)

(12)   $D_i = 100 - \dfrac{E \times N_{iE}}{F \times N_{iF}} \times 100$

But from equation (1) it is easily seen that

(13)   $\dfrac{E}{F} = \dfrac{I_F}{I_E}$

Thus, by substituting (13) into (12) it is shown that

$$(14) \quad D_i = 100 - \frac{I_F \times N_{iE}}{I_E \times N_{iF}} \times 100$$

If an external indicator (A) is administered to the animal and excreted uniformly, then fecal output (E) can be calculated from a sample of the feces ($F_s$) by the following relation

$$(15) \quad E = \frac{E_s}{\dfrac{(E_s \times A_E)}{(100)} \div A}$$

Therefore, digestibility can be calculated by substituting equation (15) into equation (4) as follows

$$(16) \quad D_{dm} = 100 - \frac{\dfrac{E_s}{\dfrac{(E_s \times A_E)}{(100)} \div A}}{F}$$

But in range and pasture trials, feed intake (F) is not regulated. However, by combining an internal indicator and an external indicator feed intake (F) may be calculated by using equation (2) as follows

$$(17) \quad F = \frac{E_s}{\dfrac{(E_s \times A_E)}{(100)} \div A} \times \frac{I_E}{I_F}$$

Once feed intake (F) and fecal output (E) have been calculated with equations (15) and (17), then the apparent digestibility of any nutrient may be calculated by equation (10).

The concentration of a digestible nutrient in the diet ($C_i$) becomes

$$(18) \quad C_i = \frac{N_i \times D_i}{100}$$

It will now be shown that microdigestion estimates for a nutrient can be used to calculate feed intake (F). For an example, microdigestion of cellulose ($M_c$) will be used. It is readily seen that the weight of cellulose in the diet ($F_c$) is

$$(19) \quad F_c = \frac{F \times N_{cF}}{100}$$

which rearranges to

$$(20) \quad F = \frac{F_c}{N_{cF}} \times 100$$

Similarly, it follows that the weight of fecal cellulose ($E_c$) is

$$(21) \quad E_c = \frac{E \times N_{cE}}{100}$$

Now, substituting into equation (20) by use of equation (9) it is seen

$$(22) \quad F = \frac{\dfrac{E_c}{100 - M_c} \times 100}{N_{cF}} \times 100$$

Then using equation (21) to calculate $E_c$ this becomes

$$(23) \quad F = \frac{\dfrac{E \times N_{cE}}{100 - M_c}}{N_{cF}} \times 100$$

which simplifies to

$$(24) \quad F = \frac{N_{cE} \times E}{100 \times N_{cF} - N_{cF} \times M_c} \times 100$$

If microdigestion of cellulose is expressed in a decimal form, then this simplifies to

$$(25) \quad F = \frac{N_{cE} \times E}{N_{cF}(1 - M_c)}$$

86

Chapter 5

## References

AGRICULTURAL BOARD: NATIONAL ACADEMY OF SCIENCES, NATIONAL RESEARCH COUNCIL. & AMERICAN SOCIETY OF RANGE MANAGEMENT (1962). Basic problems and techniques, in range research. *Nat. Res. Coun. Pub.* 890.

ALEXANDER R.A., HENTGES, JR. J.F., McCALL J.T. & Ash W.O. (1962). Comparative digestibility of nutrients in roughages by cattle and sheep. *J. Anim. Sci.* 21, 373–376.

ALKON P.U. (1961). Nutritional and acceptability values of hardwood slash as winter deer browse. *J. Wildl. Mgmt.* 25, 77–81.

ALLDEN W.G. (1962). Rate of herbage intake and grazing time in relation to herbage availability. *Aust. Soc. Anim. Prod.* 4, 163–166.

AMERICAN SOCIETY OF ANIMAL PRODUCTION (1959). *Techniques and Procedures in Animal Production Research.* Am. Soc. Anim. Prod.

ANDERSON A.E., SNYDER W.A. & BROWN G.W. (1965). Stomach content analyses related to condition in mule deer, Guadalupe Mountains, New Mexico. *J. Wildl. Mgmt.* 29, 352–356.

ANNISON E.F. & LEWIS D. (1959). *Metabolism in the Rumen.* Methuen and Co., Ltd. London, England. 184 p.

ARCHIBALD J.G., FENNER H., OWEN, JR. D.F. & BARNES H.D. (1961). Measurement of the nutritive value of alfalfa and timothy hay by varied techniques. *J. Dairy Sci.* 44, 2232–2241.

ASPLUND J.M., BERG R.T., McELROY L.W. & PIGDEN W.J. (1958). Dry matter loss and volatile fatty acid production in the artificial rumen as indices of forage quality. *Can. J. Anim. Sci.* 38, 171–180.

AXELSSON J. & KIVIMAE A. (1951). Comparison between the accuracies of the direct and indirect methods in digestion trials with wethers. *Acta. Scandanavia* 1, 282–290.

BALCH C.C. & JOHNSON V.W. (1950). Factors affecting the utilization of food by dairy cows. II. Factors influencing the rate of breakdown of cellulose (cotton thread) in the rumen of the cow. *Br. J. Nutr.* 4, 389–396.

BARNES R.J. & MOTT G.O. (1962). Comparison of *in vitro* rumen fermentation procedures. *J. Anim. Sci.* 21, 1033.

BARNETT A.J.G. & REID R.L. (1961). *Reactions in the Rumen.* Edward Arnold Publ., Ltd., London, England. 252 p.

BARTLETT J.M. (1904). Digestion experiments with sheep and steers. *Maine Ag. Exp. Sta. Bull.* 110.

BAUMGARDT B.R., CASON J.L. & TAYLOR M.W. (1962a). Evaluation of forages in the laboratory. I. Comparative accuracy of several methods. *J. Dairy Sci.* 45, 59–61.

BAUMGARDT B.R., TAYLOR M.W. & CASON J.L. (1962b). Evaluation of forages in the laboratory. II. Simplified artificial rumen procedure for obtaining repeatable estimates of forage nutritive value. *J. Dairy Sci.* 45, 62–68.

BEESTON J.W.U. & HOGAN J.P. (1960). The estimation of pasture intake by the grazing sheep. *Aust. Soc. Anim. Prod., Proc.* 3, 79–82.

BISSELL H.D. (1959). Interpreting chemical analysis of browse. *Calif. Fish Game* 45, 57–58.

BISSELL H.D., HARRIS B., STRONG H. & JAMES F. (1955). The digestibility of certain natural and artificial foods eaten by deer in California. *Calif. Fish Game* 41, 57–78.

BISSELL H.D. & WEIR W.C. (1957). The digestibilities of interior live oak and chamise by deer and sheep. *J. Anim. Sci.* **16**, 476–480.

BLAXTER K.L., McGRAHAM N. & WAINMAN F.W. (1956). Some observations on the digestibility of food by sheep, and on related problems. *Br. J. Nutr.* **10**, 69–91.

BLAXTER K.L. & MITCHELL H.H. (1948). The factorization of the protein requirements of ruminants of the protein values of feeds, with particular reference to the significance of the metabolic fecal nitrogen. *J. Anim. Sci.* **7**, 351–372.

BOWDEN D.M. & CHURCH D.C. (1962a). Artificial rumen investigations. I. Variability of dry matter and cellulose digestibility and production of volatile fatty acids. *J. Dairy Sci.* **45**, 972–979.

BOWDEN D.M. & CHURCH D.C. (1962b). Artificial rumen investigations. II. Correlations between *in vitro* and *in vivo* measures of digestibility and chemical components of forages. *J. Dairy Sci.* **45**, 980–985.

BRISSON G.J. (1960). Indicator methods for estimating amount of forage consumed by grazing animals. *Int. Grassland Congr., Proc.* **8**, 435–438.

BROWN, DOROTHY. (1954). Methods of surveying and measuring vegetation. *Commonwealth Bureau Pastures Field Crops. Bull.* 42.

BRUSVEN M.A. & MULKERN G.B. (1960). The use of epidermal characteristics for the identification of plants recovered in fragmentary condition from the crops of grasshoppers. *North Dakotas Res. Rep. No.* 3.

BURNS L.A. (1958). Digestion studies and gains of beef cattle grazing on crested wheatgrass. M.S. Thesis. Univ. Idaho. 43 p.

CABALLERO H., GALLI I.O. & MOORE J.E. (1962). A comparison of forage evaluation methods. *Am. Soc. Anim. Sci.* **54**, 1–5.

CHAMROD A.D. & BOX T.W. (1964). A point frame for sampling rumen contents. *J. Wildl. Mgmt.* **28**, 473–477.

CHIPPENDALE G. (1962). Botanical examination of kangaroo stomach contents and cattle rumen contents. *Aust. J. Sci.* **25**, 21–22.

CHURCH D.C. & PETERSON R.G. (1960). Effect of several variables on *in vitro* rumen fermentation. *J. Dairy Sci.* **43**, 81–92.

CIPOLLONI M.A., SCHNEIDER B.H., LUCAS H.L. & PAVLECH H.M. (1951). Significance of the differences in digestibility of feeds by cattle and sheep. *J. Anim. Sci.* **10**, 337–343.

CLANTON D.C. (1961). Comparison of 7- and 10-day collection periods in digestion and metabolism trials with beef heifers. *J. Anim. Sci.* **20**, 640–643.

CLARK K.W. & MOTT G.O. (1960). The dry matter digestion *in vitro* of forage crops. *Can. J. Plant Sci.* **40**, 123–129.

COOK C.W., BLAKE J.T. & McCALL J.W. (1963). Use of esophageal-fistula cannulae for collection forage samples from both sheep and cattle grazing in common. *J. Anim. Sci.* **22**, 579–581.

COOK C.W. & HARRIS L.E. (1950). The nutritive content of the grazing sheep's diet on summer and winter ranges of Utah. *Utah Agr. Exp. Sta. Bull.* 342.

COOK C.W. & HARRIS L.E. (1951). A comparison of the lignin ratio technique and the chromogen method of determining digestibility and forage consumption of desert range plants by sheep. *J. Anim. Sci.* **10**, 565–573.

COOK C.W., MATTOX J.E. & HARRIS L.E. (1961). Comparative daily consumption and digestibility of summer range forage by wet and dry ewes. *J. Anim. Sci.* **20**, 866–870.

COOK C.W. & STODDARD L.A. (1961). Nutrient intake and livestock responses on seeded foothill ranges. *J. Anim. Sci.* **20**, 36–41.

COOK C.W., STODDARD L.A. & HARRIS L.E. (1954). The nutritive value of winter range plants. *Utah Ag. Exp. Sta. Bull.* 372.

COOK C.W., STODDARD L.A. & HARRIS L.E. (1956). Comparative nutritive value and palatability of some introduced and native forage plants for spring and summer grazing. *Utah Ag. Exp. Sta. Bull.* 385.

COOK C.W., TAYLOR K. & HARRIS L.E. (1962). The effect of range condition and intensity of grazing upon daily intake and nutritive value of the diet on desert ranges. *J. Range Mgmt.* **15**, 1–6.

CORBETT J.L. (1960). Faecal-index techniques for estimation herbage consumption by grazing animals. *Int. Grassland Congr., Proc.* **8**, 438–442.

CRAMPTON E.W. & LLOYD L.E. (1959). *Fundamentals of Nutrition.* W.H. Freeman Co., San Francisco, Calif. 494 p.

DIETZ D.R., UDALL R.H. & YEAGER L.E. (1962). Chemical composition and digestibility by mule deer of selected forage species, Cache La Poudre Range, Colorado. *Colo. Dep. Game Fish Tech. Publ. No.* 14.

DIJKSTRA N.D., WEIDE H.J. & ADRICHEM P.W.M. VAN (1962). A comparison of the digestibility of roughage by wethers and by dairy cows. *Versl. landbouwk. Onderozock.* **68**, 9–21.

DRUCE E. & WILCOX J.S. (1949). The application of modified procedures in digestibility studies. *Empire J. Exp. Ag.* **17**, 188–192.

ELLIOTT R.C. & FOKKEMA K. (1961). Herbage consumption studies of beef cattle. I. Intake studies on Afrikander and Mashona cows, 1958–1959. *Rhodesian Ag. J.* **58**, 49–57.

FELS H.E., MOIR R.J. & ROSSITER R.C. (1959). Herbage intake of grazing sheep in southwestern Australia. *Aust. J. Ag. Res.* **10**, 237–247.

FINA L.R., KEITH C.L., BARTLEY E.E., HARTMEN P.A. & JACOBSON N.L. (1962). Modified *in vivo* artificial rumen (Vivar) techniques. *J. Anim. Sci.* **21**, 930–934.

FOOT, JANET Z. & ROMBERG, BARBARA. (1965). The utilization of roughage by sheep and the red kangaroo, *Macropus Rufus* (Desmarest). *Aust. J. Ag. Res.* **16**, 429–435.

FORBES E.B. & BEEGLE F.M. (1916). The mineral metabolism of the milk cow. *Ohio Ag. Exp. Sta. Bull.* 295.

FORBES E.B., BRATZLER J.W., BLACK A. & BRAMAN W.W. (1937). The digestibility of rations by cattle and sheep. *Pa. Ag. Exp. Sta. Bull.* 339.

FORBES E.B., ELLIOT R.J., SWIFT R.W., JAMES W.H. & SMITH V.F. (1946). Variation in determinations of digestive capacity of sheep. *J. Anim. Sci.* **5**, 298–305.

FORBES E.B. & GRINDLEY H.S. (1923). On the formulation of methods of experimentation in animal production. *Nat. Res. Coun.* 6 Bull. (*Part 2, No.* 33).

FORBES R.M. & GARRIGUS W.P. (1948). Application of a lignin ratio technique to the determination of the nutrient intake of grazing animals. *J. Anim. Sci.* **7**, 373–382.

FRAPS G.S. (1912). Digestion experiments with Texas hays and fodders. *Texas Ag. Exp. Sta. Bull.* 147.

FRENCH M.H. (1956). The effects of restricted intakes on the digestibility of hays by East African hair-sheep and Zebu oxen. *Empire J. Exp. Ag.* **24**, 235–244.

FRENCH C.E., MCEWEN L.C., MAGRUDER N.D., INGRAM R.H. & SWIFT R.W. (1955). Nutritional requirements of white-tailed deer for growth and antler development. *Pa. Ag. Exp. Sta. Bull.* 600.

GALLUP W.D., HOBBS C.S. & BRIDGES H.M. (1945). The use of silica as a reference substance in digestion trials with ruminants. *J. Anim. Sci.* **4**, 68–71.

GALLUP W.D. & KUHLMAN A.H. (1931). A preliminary study of the apparent digestibility of protein by modified procedures. *J. Ag. Res.* **42**, 665–669.

GARRIGUS W.P. (1934). The forage consumption of grazing steers. *Am. Soc. Anim. Prod., Proc.* **27**, 66–69.

GONCAROVA M.E., LUTFULLINA M.S., SEVEDKINA Z.G. & HARIN S.A. (1960). Estimation of consumption and digestibility of pasture herbage from inert substances. *Sborn. Nauc. Trud. Semipalatinsk. Zooteh. -vet Inst.* (1959–1960) **2**, 5–12.

GRAINGER R.B., GOY N. & BAKER F.H. (1960). Relationship of feed intake and length of collection period to apparent digestibility of a self-fed, pelleted lamb ration. *J. Anim. Sci.* **19**, 1150–1152.

GRASSLAND RESEARCH INSTITUTE (1961). Research techniques in use at the Grassland Research Institute. *Commonwealth Bureau Pastures Field Crops Bull.* 45.

GREENHALGH J.F.D. & CORBETT J.L. (1960). The indirect estimation of the digestibility of pasture herbage. I. Nitrogen and chromogen as faecal index substances. *J. Ag. Sci.* **55**, 371–376.

GREENHALGH J.F.D., CORBETT J.L. & McDONALD I. (1960). The indirect estimation of the digestibility of pasture herbage. II. Regressions of digestibility of fecal nitrogen concentrations; their determination in continuous digestibility trials and the effect of various factors on their accuracy. *J. Ag. Sci.* **55**, 377–386.

GROENWALD J.W., MYBERGH J.S. & LAWRENCE G.B. (1950). Digestibility of lucerne hay with special reference to experimental technique in digestion trials. *Onderstepoort J. Vet. Sci. Anim. Ind.* **24**, 67–85.

GUPTA B.N., MAJUMDAR B.N. & KEHAR N.D. (1962). Studies on indirect methods of determining feed digestibilities. III. Nitrogen as fecal index indicator in herbage intake determinations. *Ann. Biochem. Exp. Med.* **22**, 105–112.

HALE O.M., HUGHES R.H. & KNOX F.E. (1962). Forage intake by cattle grazing wiregrass range. *J. Range Mgmt.* **15**, 6–9.

HALLS L.K., HALE O.M. & KNOX F.E. (1957). Seasonal variation in grazing use, nutritive content, and digestibility of wiregrass forage. *Ga. Ag. Exp. Sta. Tech. Bull. N.S.* 11.

HARKER K.W., TORELL D.T. & VAN DYNE G.M. (1964). Botanical examination of forage from esophageal fistulas in cattle. *J. Anim. Sci.* **23**, 465–469.

HEADY H.F. (1964). Palatability of herbage and animal preference. *J. Range Mgmt.* **17**, 76–82.

HEADY H.F. & VAN DYNE G.M. (1965). Prediction of weight composition from point samples on clipped herbage. *J. Range Mgmt.* **18**, 144–148.

HILL K.R., REPP W.W., WATKINS W.E. & KNOX J.H. (1961). Estimation of feed intake and digestibility by 4-, 6-, and 24-hour fecal collections using lignin as the indicator with heifers. *Western Sect. Am. Soc. Anim. Prod.* **12**, 1–5.

HODGSON R.E. & KNOTT J.C. (1932). Apparent digestibility of, and nitrogen, calcium, and phosphorus balance of dairy heifers on, artificially dried pasture herbage. *J. Ag. Res.* **45**, 557–563.

HOFLUND S., QUIN J.I. & CLARK R. (1948). Studies on the alimentary tract of the Merino sheep. XV. The influence of different factors on the rate of cellulose digestion (a) in the rumen and (b) in the ruminal ingesta as studied *in vitro*. *Onderstepoort J. Vet. Sci. Anim. Ind.* **23**, 395–409.

HOLDER J.M. (1962). Supplementary feeding on grazing sheep—its effect on pasture intake. *Aust. Soc. Anim. Prod.* **4**, 154–159.

HOLMES W., JONES J.G.W. & DRAKE-BROCKMAN R.M. (1961). The feed intake of grazing cattle. II. The influence of size of animal on feed intake. *Anim. Prod.* **3**, 251–260.

HOPSON J.D., JOHNSON R.R. & DEHORITY B.A. (1963). Evaluation of the dacron bag technique as a method for measuring cellulose digestibility and rate of forage digestion. *J. Anim. Sci.* **22**, 448–453.

HOWES J.R., HENTEGES, JR. J.F. & DAVIS G.K. (1963). Comparative digestive powers of Hereford and Brahman cattle. *J. Anim. Sci.* **22**, 22–26.

HUNGATE R.E., PHILLIPS G.D., HUNGATE D.P. & MACGREGOR A. (1960). A comparison of the rumen fermentation in European and Zebu cattle. *J. Ag. Sci.* **54**, 196–201.

HUTCHINSON K.J. (1958). Factors governing fecal nitrogen wastage in sheep. *Aust. J. Ag. Res.* **9**, 508–520.

ICHHPONANI J.S., MAKKAR G.S., SIDHU G.S. & MOXON A.L. (1962). Cellulose digestion in water buffalo and Zebu cattle. *J. Anim. Sci.* **21**, 1001.

JOHNSON R.R. (1963). *In vitro* rumen fermentation techniques. *J. Anim. Sci.* **22**, 792–800.

JOHNSON R.R., DEHORITY B.A., CONRAD H.R. & DAVIS R.R. (1962). Relationship of *in vitro* cellulose digestibility of undried and dried mixed forages to their *in vivo* dry matter digestibility. *J. Dairy Sci.* **45**, 250–252.

JOINT COMMITTEE: AMERICAN SOCIETY OF AGRONOMY, AMERICAN DAIRY ASSOCIATION, AMERICAN SOCIETY OF ANIMAL PRODUCTION, AND AMERICAN SOCIETY OF RANGE MANAGEMENT (1962). *Pasture and Range Research Techniques*. Comstock Publ. Ass. Ithaca, N.Y. 242 p.

JORDAN R.M. & STAPLES G.E. (1951). Digestibility comparisons between steers and lambs fed prairie hays of different quality. *J. Anim. Sci.* **10**, 236–243.

JULANDER O., FERGUSON B.R. & DEALY J.E. (1963). Measure of animal range use by signs. P. 102–108. In: U.S. Forest Serv., Range Res. Methods, *U.S. Dept. Ag. Misc. Publ.* 940.

KAMSTRA L.D. & MILLER N.T. (1960). Ruminal variation in pH, temperature and *in vitro* activity as affected by ration change and season. *South Dakota Acad. Sci. Proc.* **39**, 84–88.

KENNEDY P.B. & DINSMORE S.C. (1909). Digestion experiments on the range. *Nevada Ag. Exp. Sta. Bull.* 71.

KERCHER C.J. (1962). Nylon bag studies. In: *Western Regional Res. Proj.* W-34, *Annu. Progress Rep., Wyoming Ag. Exp. Sta.*

KING W.A., LEE J., III, WEBB H.J. & RODERICK D.B. (1960). Comparison of 6- and 10-day collection periods for digestion trials with dairy heifers. *J. Dairy Sci.* **43**, 388–392.

KLEIBER M. (1961). *The Fire of Life—an introduction to animal energetics*. John Wiley & Sons, Inc. New York. 454 p.

KLEIN D.R. (1962). Rumen contents analysis as an index to range quality. *North Amer. Wildl. Nat. Resources Conf., Trans.* **27**, 150–164.

KNOTT J.C., MURER H.K. & HODGSON R.E. (1936). The determination of apparent digestibility of green and cured grass by modified procedures. *J. Ag. Res.* **53**, 553–556.

LAMBOURNE L.J. (1957). Measurement of feed intake of grazing sheep. II. The estimation of faeces output using markers. *J. Ag. Sci.* **48**, 415–425.

LANCASTER R.J. (1949). The measurement of feed intake of grazing cattle and sheep. I. A method of calculating the digestibility of pasture based on the nitrogen content of faeces derived from the pasture. *New Zealand J. Sci. Tech.* **A-31**, 31–38.

LANCASTER R.J. (1954). Measurement of feed intake of grazing cattle and sheep. V. Estimation of the feed-to-faeces ratio from the nitrogen content of the faeces of pasture fed cattle. *New Zealand J. Sci. Tech.* **36**, 15–20.

LEOPOLD A.S., RINEY T., McCAIN R. & TEVIS JR. L. (1951). The Jawbone deer herd. *Calif. Dept. Nat. Res. Game Bull.* 4.

LLOYD L.E., PECKHAM H.E. & CRAMPTON E.W. (1956). The effect of change of ration on the required length of preliminary feeding period in digestion trials with sheep. *J. Anim. Sci.* **15**, 846–853.

LUSK J.W., BROWNING C.B. & MILES J.T. (1962). Small sample *in vivo* cellulose digestion procedure for forage evaluation. *J. Dairy Sci.* **45**, 69–73.

LYNCH P.B. (1960). Conduct of field experiments. *New Zealand Dept. Ag. Bull. No.* 399.

MAGRUDER N.D., FRENCH C.E., McEWEN L.C. & SWIFT R.W. (1957). Nutritional requirements of white-tailed deer for growth and antler development. II. Experimental results of the third year. *Pa. Ag. Exp. Sta. Bull.* 628.

MARTIN A.C. & KORSCHGEN L.J. (1963). Food-habits procedures. P. 320–329. In: *Wildlife Investigational Techniques.* The Wildlife Society. Washington, D.C. 419.

MAYNARD L.A. & LOOSLI J.K. (1962). *Animal Nutrition.* McGraw-Hill Book Co., New York. 533 p.

MILFORD R. (1957). The usefulness of lignin as an index of forage consumed by ruminants. P. 79–85. In: *Lignification of Plants in Relation to Ruminant Nutrition.* Symp. Div. Plant Ind., Proc. CSIRO. Canberra, Australia.

MINSON D.J. & KEMP C.D. (1961). Studies in the digestibility of herbage. IX. Herbage and faecal nitrogen as indicators of herbage organic matter digestibility. *J. Br. Grassland Soc.* **16**, 76–79.

MITSUMATA M., TAKANO N., MIYASHITA A. & WATARAI H. (1959). The quantities of herbage nutrition eaten by cows on the pasture when tether-grazing. *Hokkaido Nat. Ag. Exp. Sta. Bull.* 74.

MOORE J.E., JOHNSON R.R. & DEHORITY B.A. (1962). Adaptation of an *in vitro* system to the study of starch fermentation by rumen bacteria. *J. Nutr.* **76**, 414–422.

MORRIS M.S. & SCHWARTZ J.E. (1957). Mule deer and elk food habits on the National Bison Range. *J. Wildl. Mgmt.* **21**, 189–193.

MOSBY H.S. (Ed.) (1963). *Wildlife Investigational Techniques.* The Wildlife Society, Washington, D.C. 419 p.

NICHOL A.A. (1938). Experimental feeding of deer. *Ariz. Ag. Exp. Sta. Tech. Bull.* 75.

NICHOLSON J.W.G., HAYNES E.H., WARNER R.G. & LOOSLI J.K. (1956). Digestibility of various rations by steers as influenced by the length of preliminary feeding period. *J. Anim. Sci.* **15**, 1172–1179.

NORRIS J. (1943). Botanical analysis of stomach contents as a method of determining forage consumption of range sheep. *Ecol.* **24**, 224–251.

OWEN J.B. (1961). A new method of estimating the dry matter intake of grazing sheep from their faecal output. *Nature* **192**, 92.

PHILLIPS E.A. (1959). *Methods of Vegetation Study.* Henry Holt & Co., New York. 107 p.

PLATIKNOFF N. VON & POPOFF I. (1937). Composition and nutritive value of wheat screenings; simplification of the method for determination of the digestibility and nutritive value of fodder. *Sofiisk. Univ. Agron. Lesov. Fukult. God.* **16**, 400–429.

PRITCHARD G.I., FOLKINS L.P. & PIGDEN W.J. (1963). The *in vitro* digestibility of whole grasses and their parts at progressive stages of maturity *Can. J. Plant Sci.* **43**, 79–87.

PROTASENJA T.P. (1956). New method of study of digestion of carbohydrates, proteins, and fats in animals with fistulae. *Fiziol. Z. SSR. Secenova* **42**, 420–433.

PURSER D.B. & MOIR R.J. (1959). Ruminal flora studies in the sheep. IX. The effect of pH on the ciliate population of the rumen *in vivo. Aust. J. Ag. Res.* **10**, 555–564.

QUICKE G.V., BENTLEY O.G., SCOTT H.W. & MOXON A.L. (1959). Cellulose digestibility *in vitro* as a measure of the digestibility of forage cellulose in ruminants. *J. Anim. Sci.* **18**, 275–287.

QUINN J.I., VAN DER WATH J.G. & MYBERG S. (1938). Studies on the alimentary tract of Merino sheep in South Africa. IV. Description of experimental technique. *Ondestepoort J. Vet. Sci. Anim. Ind.* **11**, 341–382.

RAYMOND W.F. (1954). Studies in the digestibility of herbage. III. The use of faecal collection and chemical analysis in pasture studies; (a) Ratio and tracer methods. *J. Br. Grassland Soc.* **9**, 61–67.

RAYMOND W.F., HARRIS C.E. & HARKER V.G. (1953). Studies on the digestibility of herbage. I. Technique of measurement of digestibility and some observations on factors affecting the accuracy of digestibility data. *J. Br. Grassland Soc.* **8**, 301–314.

RAYMOND W.F., KEMP C.D., KEMP A.W. & HARRIS C.E. (1954). Studies in the digestibility of herbage. IV. The use of faecal collection and chemical analyses in pasture studies; (b) Faecal index methods. *J. Br. Grassland Soc.* **9**, 69–82.

REID J.T. (1962). Indicator methods in herbage quality studies. P. 45–46. In: Joint Committee (1962). *Pasture and Range Research Techniques.* Comstock Publ. Ass. Ithaca, New York. 242 p.

REID J.T. & KENNEDY W.K. (1956). Measurement of forage intake by grazing animals. *Int. Grassland Congr., Proc.* **7**, 3–8.

REID J.T., WOOLFOLK P.G., HARDISON W.A., MARTIN C.M., BRUNDAGE A.L. & KAUFMANN R.W. (1952). A procedure for measuring the digestibility of pasture forage under grazing conditions. *J. Nutr.* **46**, 255–269.

SALO, MAIJA-LUSA. (1957a). Lignin studies. I. Investigations concerning lignin determination. *J. Sci. Ag. Soc. Finland* **29**, 185–193.

SALO, MAIJA-LUSA. (1957b). Lignin studies. II. The lignin content and properties of lignin in different materials. *J. Sci. Ag. Soc. Finland* **29**, 202–210.

SALO MAIJA-LUSA. (1958). Lignin studies. III. Lignin as a tracer in digestibility investigations. *J. Sci. Ag. Soc. Finland* **30**, 97–104.

SCHNEIDER B.H. (1947). *Feeds of the World: their digestibility and composition.* Jarrett Printing Co., Charleston, W. Va. 299 p.

SCHNEIDER B.H. & ELLENBERGER H.B. (1927). Apparent digestibility as affected by length of trial and by certain variations in the ration. *Vt. Ag. Exp. Sta. Bull.* 270.

SCHNEIDER B.H. & LUCAS H.L. (1950). The magnitude of certain sources of variability in digestibility data. *J. Anim. Sci.* **9**, 504–512.

SCHNEIDER B.H., SONI B.K. & HAM W.E. (1955). Methods for determining consumption and digestibility of pasture forages. *Wash. Ag. Exp. Sta. Tech. Bull.* 16.

SELL O.E., REID J.T., WOOLFOLK P.G. & WILLIAMS R.E. (1959). Inter-society forage evaluation symposium. *Agron. J.* **51**, 212–245.

SKULMOWSKI J., SZYMENSKI A. & WYSZYNSKI T. (1943). Frakische anwendung einer neven, der sogemanten 'qualtativen'oder indikator-methods zur bestimmung der verdaulich keft des futters. *Ber. Lardow. Forschunganst das General-Governments* **1**, 76.

SMART W.W.G. JR., MATRONE G., SHEPHERD W.O., HUGHES R.H. & KNOX F.E. (1960). The study of the comparative composition and digestibility of cane forage (*Arundinaria* spp.). *N.C. Ag. Exp. Sta. Tech. Bull.* 140.

SMITH A.D. (1950a). Sagebrush as a winter feed for deer. *J. Wildl. Mgmt.* **14**, 285–289.

SMITH A.D. (1950b). Feeding deer on browse species during winter. *J. Range Mgmt.* **3**, 130–132.

SMITH A.D. (1952). Digestibility of some native forages for mule deer. *J. Wildl. Mgmt.* **16**, 309–312.

SMITH A.D. (1953). Consumption of native forage species by captive mule deer during summer. *J. Range Mgmt.* **6**, 30–37.

SMITH A.D. (1957). Nutritive value of some browse plants in winter. *J. Range Mgmt.* **10**, 162–164.

SMITH A.D. (1964). Defecation rates of mule deer. *J. Wildl Mgmt.* **28**, 435–444.

SMITH A.D., TURNER R.B. & HARRIS G.A. (1956). The apparent digestibility of lignin by mule deer. *J. Range Mgmt.* **9**, 142–145.

SPARKS D.R. & MALECHEK J.C. (1967). Estimating percentage dry weights in diets using a microscopic technique. *J. Range Mgmt.* (manuscript submitted).

STAPLES G.E. & DINUSSON W.E. (1951). A comparison of the relative accuracy between seven-day and ten-day collection periods in digestion trials. *J. Anim. Sci.* **10**, 244–250.

STIELAU W.J. (1960). The length of the preliminary feeding period required in digestibility experiments with sheep. *S. Afr. J. Ag. Sci.* **3**, 433–439.

TALBOT L.M. & TALBOT, MARTHA H. (1962). Food preferences of some East African wild ungulates. *East African Ag. Forest. J.* **27**, 131–138.

TAYLOR B.G., REPP W.W. & WATKINS W.E. (1960). An artificial rumen technique versus conventional digestion trials for determining digestibility of blue grama, sudan, and alfalfa hays. *Western Sect. Am. Soc. Anim. Prod., Proc.* **11**, 1–5.

TILLEY J.M.A. & TERRY R.A. (1963). A two-stage technique for the *in vitro* digestion of forage crops. *J. Br. Grassland Soc.* **18**, 104–111.

TOPPS J.H. (1962). Studies of natural herbage of the sub-tropics. I. The digestibility of herbage grazed by cattle. *J. Ag. Sci.* **58**, 387–391.

ULLREY D.E., YOUATT W.G., JOHNSON H.E., KU P.K. & FAY L.D. (1964). Digestibility of cedar and aspen browse for the white-tailed deer. *J. Wildl. Mgmt.* **28**, 791–797.

U.S.D.A. FOREST SERVICE (1959). Techniques and methods of measuring understory vegetation (Symp.) Southern Forest Exp. Sta. and South-eastern Forest Exp. Sta.

VAN DYNE G.M. (1960). Range investigations. In: *Western Regional Res. Proj. W-34, Ann. Progress Rep., Mont. Ag. Exp. Sta.* (p. 9–60).

VAN DYNE G.M. (1962). Micro-methods for nutritive evaluation of range forages *J. Range Mgmt.* **15**, 303–314.

VAN DYNE G.M. (1963). An artificial rumen system for range nutrition studies. *J. Range Mgmt.* **16**, 146–147.

VAN DYNE G.M. (1967). Variables affecting *in vitro* rumen fermentation studies in forage evaluation: a review. *J. Anim. Sci.* (manuscript submitted, 54 p.).

VAN DYNE G.M. & HEADY H.F. (1965a). Botanical composition of sheep and cattle diets on a mature annual range. *Hilgardia* **36**, 465–492.

VAN DYNE G.M. & HEADY H.F. (1965b). Dietary chemical composition of cattle and sheep grazing in common on a dry annual range. *J. Range Mgmt.* **18**, 78–85.

VAN DYNE G.M. & LOFGREEN G.P. (1964). Comparative digestion of dry annual range forage by cattle and sheep. *J. Anim. Sci.* **23**, 823–832.

VAN DYNE G.M. & MEYER J.H. (1964a). A method for measurement of forage intake of grazing livestock using microdigestion techniques. *J. Range Mgmt.* **17**, 204–208.

VAN DYNE G.M. & MEYER J.H. (1964b). Forage intake by cattle and sheep on dry annual range. *J. Anim. Sci.* **23**, 1108–1115.

VAN DYNE G.M. & TORELL D.F. (1964). Development and use of esophageal fistula: a review. *J. Range Mgmt.* **17**, 7–19.

VAN DYNE G.M. & WEIR W.C. (1964). Variations among cattle and sheep in digestive power measured by microdigestion techniques. *J. Anim. Sci.* **23**, 1116–1123.

VAN DYNE G.M. & WEIR W.C. (1966). Comparison of microdigestion techniques under range and drylot conditions. *J. Ag. Sci.* **67**, 381–387.

VAN KEURAN R.W. & HEINEMANN W.W. (1962). Study of a nylon bag technique for *in vivo* estimation of forage digestibility. *J. Anim. Sci.* **21**, 340–345.

VERCOE J.E. & PEACE G.R. (1962). The estimation of intake of grazing sheep. I. Establishment of faecal nitrogen regressions. *J. Ag. Sci.* **59**, 343–348.

VERCOE J.E., TRIBE D.E. & PEACE G.R. (1962). Herbage as a source of digestible organic matter and digestible nitrogen for grazing sheep. *Aust. J. Ag. Res.* **12**, 689–695.

WALLMO O.C., JACKSON A.W., HAILEY T.L. & CARLISLE R.L. (1962). The influence of rain on the count of deer pellet groups. *J. Wildl. Mgmt.* **26**, 50–55.

WEBB W.L. (1959). Summer browse preferences of Adirondack white-tailed deer. *J. Wildl. Mgmt.* **23**, 455–456.

WATSON C.J., DAVIDSON W.M., KENNEDY J.W., ROBINSON C.H. & MUIR G.W. (1948). Digestibility studies with ruminants. XII. The comparative digestive powers of sheep and steers. *Sci. Ag.* **28**, 357–374.

WESTON R.H. (1959). The efficiency of wool production of grazing Merino sheep. *Aust. J. Ag. Res.* **10**, 865–885.

WHEELER R.R. (1962). Evaluation of various indicator techniques in estimating forage intake and digestibility by range cattle. Ph. D. Thesis. Ore. St. Univ., Corvallis, Ore. 85 p.

WILDT H. (1874). Uber die resorption und secretion der nehrungsbestandtheile in verdaing skanal of schafes. *J. Landev. Scaft.* **22**, 1.

WILLIAMS V.J. & CHRISTIAN K.R. (1956). Rumen studies in sheep. I. Variation in rumen microbial end-products in free-grazing sheep. *New Zealand J. Sci. Tech.* **A38**, 194–200.

# 6

# Metabolism

HENRY L SHORT* and FRANK B GOLLEY

The study of energy flow in large herbivores is confined mainly to determination of the efficiency with which range vegetation is utilized. Measurement of forage utilization or digestion is affected by the tractability of the herbivores and the technology available. A further complication to energy flow studies is that the flow rate from the plant community to the individual herbivore may vary throughout the year, because the energy requirements for maintenance and production vary. The different measures of food utilization, the energy requirements of herbivores under basal conditions and for growth and maintenance, as well as procedures for measuring metabolism, are described in the present chapter.

## Measurements of Food Utilization

**Energy.** The gross energy of a foodstuff, its heat of combustion, can be measured in a bomb calorimeter. Oxygen bomb calorimetry is described in many texts of animal nutrition (e.g., Maynard and Loosli, 1956). It essentially measures the potential energy an animal would receive from a foodstuff if the food were completely digested.

Digestible energy is the gross energy of a foodstuff minus the gross energy of the feces. It is the Total Digestible Nutrient (T.D.N.), which is calculated in standard digestion trials, and represents the first approximation of animal use. It is also the usual measure of energy flow used in ecology. Procedures for performing digestion trials with herbivores are described in the preceding chapter. Only the most tractable individuals of wild species can be used in forage digestibility studies, since the animals are usually kept in cages and

* On the staff of the Wildlife Habitat and Silviculture Laboratory which is maintained at Nacogdoches, Texas, by the Southern Forest Experiment Station in cooperation with Stephen F. Austin State College.

pens where food intake and fecal production can be accurately measured. Fecal digestion harnesses have also been used to collect samples of the total fecal production.

Metabolizable energy in herbivores represents the difference between the gross energy of a feed and the gross energy of feces, urine, and gases from fermentative processes within the gastro-intestinal tract. Urine can be quantitatively collected in a properly designed digestion stall. The metabolizable energy in the gases of fermentation is measured by procedures that will be described. The advantage of measuring metabolizable energy is that one can account for certain high-energy volatile substances which may be ingested but excreted, mostly in the urine. Such substances are included in totals for metabolizable energy but not in those for digestible energy. In ruminants, about 10% of the digestible energy is generally lost in urine, and 8% as methane (Byerly, 1967). The technical difficulties in obtaining urine and gas samples have minimized the number of studies in which the metabolizable energy of forage for herbivores has been determined.

Net energy is the most refined measure of the usefulness of food to the herbivore. It represents the remainder after metabolic energy (fecal, urinary, combustible gases of fermentation) and heat losses due to nutrient utilization have been deducted from the gross energy of a feed. The net energy is that which is available for maintenance, work, or secondary production. Its measurement provides the most precise information on how a forage is used by a particular individual at a particular time. However, this information can rarely be expected from studies of wild herbivores, because its measure incorporates all of the procedural difficulties of calculating metabolic energy and basal metabolism.

The relative sizes of the different measures of energy utilization of three common foodstuffs of nearly equal gross energy are shown in a tabulation adapted from Morrison (1950):

Percent of gross energy available to cattle

|  | Gross energy | Digestive energy | Metabolizable energy | Net energy |
|---|---|---|---|---|
| Ground corn | 100·0 | 88·2 | 74·9 | 46·0 |
| Timothy hay | 100·0 | 52·4 | 41·0 | 23·7 |
| Wheat straw | 100·0 | 41·8 | 31·1 | 5·5 |

The net energy of the ground corn was twice that of the timothy hay and eight times that of the wheat straw. With increasing fiber content, forage digestibility diminishes, and increasing proportions of the 'heat increment' (the increase in heat production following and incident to food ingestion) are accounted for by the energy expenditure required in the digestive processes. Tnis requirement may be even larger for other species of herbivores than for cattle.

Net energy may be measured in two general ways (Flatt, 1965). First, energy retention may be determined as metabolism plus animal productivity (weight gained, other tissue produced, change in body composition) under different feeding regimes. Second, eneigy retention may be determined indirectly by measuring energy losses of feces, urine, and combustible gases, and subtracting these from intake. The energy cost for maintenance and activity will be discussed in this chapter; secondary production is the topic of the next chapter. The energy expended for maintenance and activity includes a relatively invariable energy requirement (basal metabolism) and a highly variable requirement (activity and production).

### Basal Metabolism

The basal metabolism of homeothermic animals is that energy required for such normal body functions as respiration, blood circulation, maintenance of muscle tone, and the manufacture of internal secretions. It is the irreducible caloric requirement of the animal, measured in a thermally neutral environment, in the postabsorptive state, and without any stimulatory effect of food.

Energy exchange has been measured directly in a number of terrestrial vertebrates, but the instrumentation required to do so accurately is elaborate and costly. More commonly, energy exchange is measured as the respiratory exchange of oxygen and/or carbon dioxide, which is proportional to heat production under specific conditions. Indeed, Brody (1964) states that for most purposes—that is, when reactions are not endothermic or partly anaerobic, and the caloric equivalent of $O_2$ is known—measurement of $O_2$ consumption gives more reliable results than does a direct measure of heat production. Calculation of energy exchange from respiratory exchange depends on established metabolic constants. For precise measurements, oxygen consumption, carbon dioxide production, and urinary nitrogen excretion must be determined. When the amount of urinary nitrogen is known, the energy from protein oxidation can be calculated (Brody, 1964). The remaining energy exchange originates from oxidation of fat and carbohydrates.

The procedure is commonly simplified by disregarding protein metabolism entirely, measuring oxygen consumption only, and assuming an intermediate basal respiratory quotient (the volumetric ratio of $CO_2$ to $O_2$) of 0·82. At this respiratory quotient, consumption of one liter of oxygen is equivalent to 4·825 kcal. Mitchell (1962, p. 13) shows that this simplification results in an estimate of heat production about 2% lower than that made by considering protein and nonprotein metabolism in fasting animals.

Metabolism of terrestrial vertebrate populations under free living conditions can be measured directly or can be extrapolated from laboratory measurements. Direct methods of measuring metabolism under field conditions are potentially of greatest importance, and promising techniques will be discussed briefly. Laboratory methods are well known and will not be described; however, the basis for extrapolation from the laboratory to the field will be discussed.

**Direct methods of measuring metabolism in the field.** Only recently have ecologists attempted to measure metabolism directly under field conditions, and the methods have not been used extensively or tested on a variety of animals in different environments. Thus, there is little information about the relative accuracy or applicability of the methods to vertebrate populations in general. One of the most promising techniques is the $D_2{}^{18}O$ method developed by Lifson, *et al.*, (1955) for laboratory mice. LeFebvre (1962 and 1964) used the $D_2{}^{18}O$ technique to measure energy metabolism of the pigeon (*Columba livia*) in flight. Other direct methods utilize radioactive isotopes as activity minotors. Also, miniature radios have been employed to transmit information about the physiological state of the free living organism.

*The $D_2{}^{18}O$ method.* The rationale of the $D_2{}^{18}O$ method is that the oxygen of respiratory $CO_2$ is in isotopic equilibrium with the oxygen of the body water (Lifson, *et al.*, 1949). The hydrogen of the body water is lost mainly in water while the oxygen is lost both in water and in the $CO_2$ produced in respiration. Therefore, the turnover rate of body-water oxygen is significantly higher than that of hydrogen; and the difference in the two turnover rates is proportional to the $CO_2$ produced.

The two turnover rates may be determined by labelling these two components of the body water with stable isotopes of hydrogen (deuterium) and oxygen ($^{18}$Oxygen). Where the animal is given isotopes and then allowed to drink unlabelled water during the experimental period, the isotopic turnover rate is calculated from the change in specific activity of blood-water or body-water samples. Isotope specific activity is measured in a mass spectrometer.

It is necessary to validate the $D_2$ $^{18}$O method in the laboratory before it is used on free-living animals. LeFebvre (1962) compared $CO_2$ measurements by the Van Slyke manometric method with those by the $D_2$ $^{18}$O method. The difference between the two estimates was within about 10%.

*Use of radioactive isotopes to measure field matabolism of terrestrial vertebrates.* Odum and Golley (1963) discussed the use of radionuclides as metabolic indicators under field conditions. Rate of excretion of $^{65}$Zinc fed in microcurie quantities was related to important ecological variables, such as temperature, food consumption, and reproduction, in several kinds of animals. Since $^{65}$Zinc is excreted by the digestive tract, its biological half life is related directly to the activity of the system, and it was suggested that $^{65}$Zinc might prove a useful index of trophic activity and the rate of metabolism. Further, it has been shown that there is a significant relationship between $^{65}$Zinc excretion and body weight in mammals (Richard, *et al.*, 1962; Golley, *et al.*, 1965), when specimens ranging over a broad spectrum of weights are compared. Within a narrow range of weights, however, rate of excretion of $^{65}$Zinc may vary widely. Zinc excretion and uptake may be influenced by the chemical form of the metal, the quantity in the diet, the level of other elements in the diet, the physiological condition of the animal, and the mode of administration.

Several investigators at the University of Georgia have studied the relationship between $^{65}$Zinc excretion and metabolism in small mammals. $^{65}$Zinc was administered on food or by gavage at a dose of about 2 microcuries of $^{65}$Zn $Cl_2$ per animal. Subjects were then examined at daily intervals in a whole-body counter. Golley and Wiegert (1963) compared $^{65}$Zinc excretion and $O_2$ consumption, using a manometric respirometer (MacLagan and Sheahan, 1950), in two species of rodents, the meadow vole (*Microtus pennsylvanicus*) and the cotton rat (*Sigmodon hispidus*). Under the experimental conditions, a correlation between metabolism and excretion of $^{65}$Zinc could not be demonstrated. Orr (1964) measured $^{65}$Zinc excretion in cotton rats under laboratory and field conditions. In November and

December the loss rate of the laboratory group was less than that of the group living under natural conditions in an enclosure. The biological half life was 43 days in the laboratory group and 28 days in the field group, meaning that $^{65}$Zinc excretion under field conditions was 1·5 times that in the laboratory. However, in the period from March to May, no significant difference was observed in $^{65}$Zinc excretion of two other groups of animals held in the laboratory and in the field. Pulliam and Odum (1966) have recently demonstrated a relation between $^{65}$Zinc excretion and oxygen consumption in house mice (*Mus musculus*). While there was considerable variability between individuals within a temperature treatment, the correlation between respiration and biological half life was significant.

*Telemetry*. Various methods of automatically recording and transmitting physiological data from vertebrates have been developed in the laboratory and employed under field conditions. For example, Folk (1964a), utilizing a very small radio capsule with electrodes which pick up cardiac impulses from the body, has studied heart rate patterns in wild mammals. The radio capsule consists of: (1) a single transistor oscillator which produces the radio signal, (2) a transistor amplifier which amplifies voltages received by the electrodes, and (3) a voltage-sensitive capacitor which modulates the oscillator signal with the output of the amplifier. Radio capsules are placed under the skin or are worn externally, with the electrodes inserted in the skin. The signal from the radio capsule is received by a closed loop antenna connected to the receiver, which demodulates the signal. The output is then applied to a standard electrocardiograph. Using this apparatus, Folk (1964b) and Folk and Hedge (1964) found ratios of active to resting heart rates of between 1·3 and 3·5 to 1 in 8 species ranging in size from the ground squirrel to the wolf. The greatest ratio was observed in the arctic fox (*Alopex lagopus*). A similar approach has been used by Bligh and Harthoorn (1965), Bligh, *et al.*, (1965), and Rawson, *et al.*, (1965) to measure body temperature of unrestrained Welsh mountain sheep and various African animals in fenced paddocks.

**Indirect methods of measuring energy metabolism.** In the absence of suitable techniques to measure metabolism of free-living vertebrates directly, most investigators have measured the metabolic rate of confined animals in the laboratory, then extrapolated from laboratory to field environmental conditions. Laboratory methods of measuring metabolism are well known (e.g.,

Brody, 1964; and Swift and French, 1954), and need not be discussed in detail here. Since they require a respirometer or similar device, they are usually unsuitable for field use.

In basal metabolism studies, the quantity of heat produced by small animals is greater per unit of body weight than that produced by larger animals (Table 6.1). Empirical generalizations suggest that the basal metabolism of adult homeotherms, on a daily basis, can be approximated as:

$$\text{BASAL METABOLISM (kcal)} = 70 \ W^{3/4}$$

where W is equal to body weight in kg. In Table 6.1 the observed heat production (except for the elephant), expressed as a function of kg $^{0.73}$, which is very similar to $kg^{3/4}$, averages about 70. The value 70 $kcal/kg^{3/4}$ is sometimes called the interspecific mean basal metabolic rate.

The basal metabolism of large herbivores can be measured with tractable representatives of wild species, but few such trials have been accomplished. Silver *et al.* (1959) measured the basal metabolism of four white-tailed deer (*Odocoileus virginianus*), and they describe the difficulties associated with such studies of wild species. Often, metabolism is determined when the

TABLE 6.1. Heat production of 10 herbivores.*

| Animal | Kg. body wt. | Heat production per 24 hrs. per | | |
|--------|------|-------------------|-----|-----|
| | | Animal (observed) | Kg. | Kg.$^{0.73}$ |
| Albino mouse | 0·021 | 3·6 | 171 | 60 |
| Rat | ·400 | 33·2 | 83 | 65 |
| Rabbit | 2·6 | 117 | 45 | 58 |
| Goat (doe) | 36·0 | 800 | 22 | 58 |
| Sheep | 45·0 | 1160 | 26 | 72 |
| White-tailed deer** | 51·2 | 1323 | 26 | 75 |
| Small horse | 253 | 4588 | 18 | 81 |
| Cow and Steer | 500 | 6200 | 12 | 66 |
| Large horse | 703 | 11895 | 17 | 99 |
| Elephant | 3672 | 49000 | 13 | 122 |

* Data except for white-tailed deer are from Brody (1964).
** Data from Silver and Colovos (1959).

animal is fed but resting in a thermoneutral environment. The result is greater than basal metabolism, because it includes the heat increment due to feeding.

### Energy requirements for maintenance and production

The energy intake required to balance the expenditure for basal or resting metabolism is a first approximation of the maintenance requirement of herbivores. However, this value is clearly too conservative because range herbivores are both active and productive.

Blaxter (1962) has commented on the energy demands above the basal level which range cattle and sheep encounter and which are associated with eating, drinking, and moving. A 500 kg steer with a basal metabolism of 8000 kcal expends 50 kcal more energy per hr standing than lying, and each time he stands and reclines there is an additional expenditure of 12 kcal. Restlessness has a small energy cost. Young animals at play may require energy at a rate 10% above the fasting level. A foraging sheep that walks 6400 m and ascends 100 m in a day's activity may require energy at a rate 20% above the basal level. In Australia and New Zealand, grazing sheep may have requirements as great as 77% in excess of the indoor requirements (Grimes, 1966). Cattle in normal range activities may require 15% more energy than that demanded under fasting conditions. Perhaps, the energy demands of wild herbivores under usual range and normal climatic conditions are at similar heights above the fasting level.

From the above findings, Blaxter generalized that maintenance requirements of ruminants can be estimated as 1·36 kcal of metabolizable energy/kcal of fasting metabolism, or as about 1·7 kcal of apparently digested energy/kcal of fasting energy metabolism. Using the interspecies mean of 70 kcal/kg$^{3/4}$/24 hrs. and Blaxter's conversion factors, the estimated maintenance requirements for ruminants are 95·2 kcal ME/kg$^{34}$/24 hrs or 119·0 kcal DE/kg$^{3/4}$/24 hrs.

Fattening, growth, pregnancy, or lactation are factors of production that affect both food and energy consumption above the maintenance level an the efficiency of energy utilization.

The effiiciency with which ingested feeds are converted into fat (i.e., the kcal energy retained/kcal of additional metabolizable energy above maintenance) varies with the quality of the consumed food. It is lowest for feeds with a high roughage content (Blaxter, 1962). The efficiency with which energy from food is used for growth is greater in young animals than in adults. During lactation, major changes occur in the way in which food energy is utilized. During the early lactation period of high milk-producing dairy cattle fed at *ad libitum* levels, depot body fat is frequently converted to milk production (Flatt, 1966). At later stages of lactation, less of the ingested gross energy is utilized for milk production and more is mobilized to restore the depleted body reserves.

Relatively few studies of the energy metabolism of domestic ruminants at different levels of productivity have been conducted (Flatt and Coppock, 1965; Flatt, 1966). Few, if any, such energy metabolism studies have been undertaken with wild herbivores.

Severe physical activity may require great energy expenditures, and a point is reached during stress periods when more energy is spent in the search for food than can be gained from its digestion. Body heat loss varies throughout the year, as does coat insulation. Combinations of wind, hair wetness, air temperature, and humidity modify heat loss and, therefore, the metabolism required to provide necessary body warmth (Blaxter, 1962). During deep cold in New Hampshire, white-tailed deer sought shelter, remained motionless, erected coat hair, and ate little, thereby conserving heat even at the expense of food intake (Silver and Colovos, 1957).

### References

BLAXTER K.L. (1962). *The Energy Metabolism of Ruminants*. Hutchinson Scientific and Technical. London. 329 p.

BLIGH J. & HARTHOORN A.M. (1965). Continuous radiotelemetric records of the deep body temperature of some unrestrained African mammals under near-natural conditions. *J. Physiol.* **176**, 145–162.

BLIGH J., INGRAM D.L., DEYNES R.D. & ROBINSON S.G. (1965). The deep body temperature of an unrestrained Welsh Mountain Sheep recorded by a radiotelemetric technique during a 12-month period. *J. Physiol.* **176**, 136–144.

BRODY S. (1964). *Bioenergetics and Growth*. Hafner Publ. Co. Inc. N.Y. 1023 p.

BYERLY T.C. (1967). Efficiency of feed conversion. *Sci.* **157**, 890–895.

FLATT W.P. (1965). Energy values of feeds—how should they be expressed? *Proc. Wash. Anim. Nutr. Conf. Bellevue, Wash.* **18**, 13–29.

FLATT W.P. (1966). Energy metabolism results with lactating dairy cows. *J. Dairy Sci.* **49**, 230–237.

FLATT W.P. & COPPOCK C.E. (1965). Physiological factors influencing the energy metabolism of ruminants. P. 240–253, in R.W. Dougherty (Editor-in-chief), *Physiology of Digestion in the Ruminant*. Butterworths, Wash. D.C. 480 p.

FOLK E.G. JR. (1964a). The problem of electrodes for use with electrocardiograph radio capsules. *Biomedical Sciences Instrumentation*. **2**, 235–265. Plenum Press, N.Y.

FOLK E.G. JR. (1964b). Daily physiological rhythms of carnivores exposed to extreme changes in arctic daylight. *Fed. Proc.* **23**, 1221–1228.

FOLK E.G. JR. & HEDGE R.S. (1964). Comparative physiology of heart rate of unrestrained mammals. *Am. Zool.* **4**, 111.

GOLLEY F.B. (1968). *Methods of Measuring Secondary Productivity in Terrestrial Vertebrate Populations*. Working meeting on terrestrial secondary productivity, Warsaw. 43 p.

GOLLEY F.B. & WEIGERT R.G. (1963). Retention of Zinc-65 in wild small mammals. *ASB Bull.* **10**, 28.

GOLLEY F.B., WIEGERT R.G. & WALTER R.W. (1965). Excretion of orally administered Zinc-65 by wild small mammals. *Health Phys.* **11**, 719–722.

GRIMES R.C. (1966). An estimate of the energy required for maintenance and live-weight gain by young grazing sheep. *J. Ag. Sci.* **66**, 211–215.

LEFEBVRE E.A. (1962). Energy metabolism in the pigeon (*Columba livia*) at rest and in flight. Ph.D. Thesis Univ. Minn., Minneapolis. 85 p.

LEFEBVRE E.A. (1964). The use of $D_2O^{18}$ for measuring energy metabolism in *Columba livia* at rest and in flight. *Auk.* **81**, 403–416.

LIFSON N., GORDON G.B. & MCCLINTOCK R. (1955). Measurement of total carbon dioxide production by means of $D_2O^{18}$. *J. Appl. Physiol.* **7**, 704–710.

LIFSON N.G., GORDON G.B., VISSCHER M.B. & NIER A.O. (1949). The fate of utilized molecular oxygen and the source of the oxygen of respiratory carbon dioxide, studied with the aid of heavy oxygen. *J. Biol. Chem.* **180**, 803–811.

MCLAGEN N.F. & SHEAHAN M.M. (1950). The measurement of oxygen consumption in small mammals by a closed circuit method. *J. Endocrin.* **6**, 456–462.

MAYNARD L.A. & LOOSLI J.K. (1956). *Animal Nutrition*. McGraw Hill Book Co., Inc. N.Y. 484 p.

MITCHELL H.H. (1962). *Comparative Nutrition of Man and Domestic Animals*, Vol. 1. Academic Press, N.Y. 701 p.

MORRISON F.B. (1950). *Feeds and Feeding*. Morrison Publ. Co., Ithaca, N.Y. 1207 p.

ODUM E.P. & GOLLEY F.B. (1963). Radioactive tracers as an aid to the measurement of energy flow at the population level in nature. In: *Radioecology*. ed. V. Schultz and A.W. Klement, Jr. Reinhold Publ. Corp., N.Y. p. 403–410.

ORR H. (1964). Uptake and loss of Zinc-65 by the cotton rat. *Ann. Rept. to AEC, Contract At/38-1/-310*. p. 50–53.

PULLIAM R. & ODUM E.P. (1966). Correlation between 65 Zinc bioelimination and respiration rate in *Mus musculus* BALBC. *Ann. Rept. to AEC, Contract At/38-1/-310*. p. 104–106.

RAWSON R.O., STOLWIJK J.J., GRAICHEN H. & ABRAMS R. (1965). Continuous radio telemetry of hypothalamic temperatures from unrestrained animals. *J. Appl. Physiol.* **20**, 321–325.

RICHMOND C.R., FURCHNER J.E., TRAFTON G.A. & LANGHAM W.H. (1962). Comparative metabolism of radionuclides in mammals. 1. Uptake and retention of orally administered Zn[65] by four mammalian species. *Health Phys.* **8**, 481–489.

SILVER, HELENETTE & COLOVOS N.F. (1957). Native evaluation of some forage rations of deer. *N.H. Fish and Game Comm. Tech. Cir.* 15. 56 p.

SILVER, HELENETTE, COLOVOS N.F. & HAYES H.H. (1959). Basal metabolism of white-tailed deer—a pilot study. *J. Wildl. Mgmt.* **23**, 434–438.

SWIFT R.W. & FRENCH C.E. (1954). *Energy Metabolism and Nutrition.* Scarecrow Press, Wash. D.C. 264 p.

# 7

# Post Natal Growth

A J Wood and I McT Cowan

The energy balance that exists between a particular environment and the herbivore population that it supports can be evaluated in a number of ways:

(1) Productivity in terms of animal units produced per unit of time
(2) Standing crop of the particular species per unit of land area
(3) Rate of growth and age of maturity of wild living individuals
(4) Yield of edible meat on a per animal, per unit of feed consumed, on per unit of time basis
(5) Comparison of wild taken individuals with others of the same species that have been reared under captive circumstances where energy intake has been measured accurately.

A great many physical measurements commend themselves for use in the evaluation of a herbivore population. Most of these have undergone extensive use in the hands of students of domestic animal growth. It is obvious that no single criterion will be acceptable to all circumstances. The extent to which any quantitative parameters are measured will depend, in large measure, upon the resources available to a particular investigator and on the physical difficulties inherent in the application of some of the more sophisticated techniques.

It is our purpose to suggest certain philosophies and methods of approach that can yield sensitive ways of monitoring the energy inflow from the environment to the wild living ungulate. We have found that this result can be achieved by comparing the various parameters of growth taken from wild living individuals that have been reared on known planes of energy intake. Using such planes of nutrition, it is possible to rear individuals at growth rates ranging from zero to the upper genetic limit of the species. A great many sequential measurements, both physical and biochemical, can readily be obtained on such captive individuals thus providing 'standard values' as base lines for comparable measurements carried out on wild individuals in the field. If the nutrient intake required to produce a given result is known, it is then possible to extrapolate to the probable intake that occurred in the field to produce a corresponding result.

**Review of methods.** It is convenient to divide the methods that have been used for the study of post natal growth into two categories.

1. *Physical measurements.* Most studies of post natal growth have involved the determination of body weight under definable circumstances. It is also not uncommon to find various linear measurements taken to give quantitative expression to changes in physical shape and form. The published works of the Cambridge School, under the guidance of the late Sir John Hammond, on the post natal growth of the domestic meat producing animals provide useful models with which to approach the study of the wild herbivores. The procedures they have evolved and the pertinent references are given in Progress in the Physiology of Farm Animals, edited by Hammond (1954–5).

In their work they have established in clear terms the priorities which the various tissue systems possess for the utilization of ingested energy. Basically, they suggest that the tissue system which is making the most rapid growth at a particular point in time is likely to be most influenced by the energy supply that is available at that time. For example, if energy and the nutrients essential for the utilization of that energy are limiting at the time the late developing hind quarters are in their phase of maximum growth rate, then one may expect to find sub-normal development in that region.

More recent work (Wood, 1964) suggests that the ratio of the other nutrients to the energy supplied per unit weight of ration can have dramatic effects on the post natal growth rate and on the chemical composition of the growing animal. If the level of digestible energy is excessive in relation to the protein supply (both quantity and quality) of the ration, the individual consuming the ration will tend to grow slowly and will yield a carcass that is high in fat. In the converse situation, growth will again be slow but the carcass produced will be lean. Any series of gradations between extreme fatness or leanness can be achieved by manipulation of the protein to energy ratio. This work suggests that data on the protein content of wild forages is of limited value unless information on the available energy of such forages is also recorded. In too many cases the supplementation of energy marginal rations with proteins has been suggested as a means of improving growth rate. All too frequently the expensive supplementary protein is serving as an energy source rather than as a source of amino acids for the development of cellular systems. The significance of a consideration of these factors in any measures used for the expression of post natal growth is apparent.

The influence of previous nutritional history on the subsequent pattern of growth under favorable nutritional regimens has been reviewed by Wilson and Osbourn (1960), and by Winchester (1964). The inherent ability of all mammals to exhibit compensatory growth under favorable conditions suggests that the establishment of 'normal growth curves' under wild field conditions is likely to be extremely difficult, if not impossible. The rapid changes in body weight that occur during compensatory growth suggest the need to carry out frequent weight determinations on individual animals. The inaccessibility of individual wild animals for repeated weighings makes it extremely difficult to achieve this objective. Even in those cases where the animals are accessible there is the ever present concern that the events, coincident with repeated capture and measurement, may have a greater effect on growth rate than the nutritional regimen provided by the particular environment. For our captive animals we have noted that a great many factors associated with their handling and management can influence their weight status at the time of weighing and, in some cases, their weight at a subsequent weighing. The post natal growth progression can be delayed a measurable amount by coarse handling during such procedures as hoof trimming, blood sampling or even the change to a new batch of an otherwise standardized feed. As an alternative to repeated single determinations on individual wild animals, it may be more useful to establish mean values on a number of individuals taken at each time in the chronological progression of events. Such mean values can then be compared with corresponding values obtained from captive specimens. The age of the wild individuals can be assessed from studies of the rate of addition of increments to the tooth cementum established with samples taken from captive individuals of known age (Cowan and Low, 1963). In the evaluation of the wild range energy status, our experience suggests that the wild sample should be drawn from the age group 12 to 24 months. Within this age range the environment has had its opportunity to express its effects on the parameters measured, and the slow forward progression of biological events has not been able to mask earlier evidence of inimical effects.

Studies in our laboratory, carried on over the past fifteen years with deer, moose, caribou, and big horn sheep, confirm that it is essential that the selected growth parameters be measured at frequent intervals and under known nutritional conditions. Figure 7.1, taken from the Doctorate disserta-

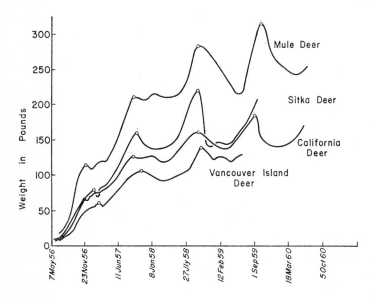

Figure 7.1 The seasonal growth of high plane males from four races of Black-tailed deer.

tion of Bandy (1965), illustrates the type of growth curve that has been established for the black-tailed deer. The mathematical procedures that have been suggested for the quantitative expression of post natal growth leave much to be desired when they are applied to curves of this type (Wood *et al.*, 1962). The complexity of the problem is further substantiated by Figure 7.2 which illustrates the pattern of feed intake self-selected by these deer when provided with a nutritionally adequate ration on an *ad libitum* basis. These growth and feed intake curves establish beyond question the need for frequent weight determinations under known conditions of feed intake. The simple linear parameters of growth in body size and proportions such as hind foot length, head length, head width, etc., should also be obtained as frequently as possible. In this way suitable regression lines can be calculated that allow the prediction of the growth sequence that will occur under defined circumstances. For example, in deer less than two years of age, it is possible, from a knowledge of hind foot length and heart girth, to make a reasonable evaluation of the range conditions on which the measured individuals were reared. (Bandy, Cowan and Wood, 1956.)

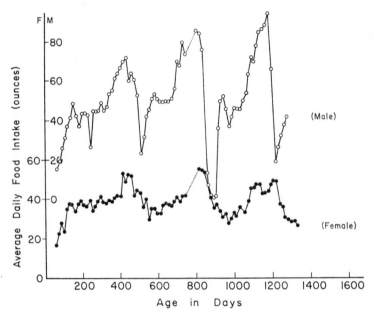

Figure 7.2. Average daily food intake in 15 day intervals for male and female deer.

From our work to date it would appear that studies should be carried out simultaneously on both sexes and in the case of the female, on both bred and unbred individuals. The stresses associated with gestation and lactation have definite effect on the growth progression. Where conditions preclude work with both sexes it is likely that the male should be selected for study because of his inherently greater absolute growth rate and because growth continues for a much longer period of time. Both of these facts suggest that the energy supply from the environment is likely to have more dramatic and more lasting effects with various post natal growth parameters in the male.

When our deer have been reared on a low plane of nutrition, the cyclic weight pattern shown in Figure 1 is greatly modulated so that one might be tempted to believe that our animals were following the S-shaped growth curve that hitherto has been considered to be normal. It is apparent that valid comparisons of productivity on an intra or inter species basis should only be made with a full knowledge of the 'ideal growth curve' for each species.

The growth pattern described above for the deer is also followed by the other wild herbivores we have studied. These include the barren ground caribou, the moose and the big horn sheep. It should be recorded that this pattern of growth can only be fully assessed under the most favorable nutritional conditions. If nutrient intake is less than ideal, growth will occur at a slower rate and the animal will not be afforded the opportunity to lay down the large fat depots that it can acquire during the periods of rapid weight accretion. The presence of these depots as reserves of energy when the animal reduces its feed intake appear to be essential to the expression of this type of growth curve.

2. *Mathematical procedures for the expression of post natal growth.* A vast literature exists on the opinions of the various investigators who have attempted to derive suitable mathematical expressions to describe the post natal growth curve. Huxley's (1932) 'Problems of Relative Growth may be cited as an example. Thompson's (1942) 'On Growth and Form' and the more recent shorter version edited by Bonner (Thompson, 1961) provide useful background material to aid in the selection of suitable expressions. Brody (1945) has reviewed and critically evaluated the suitability of a number of the mathematical procedures that have been suggested. The symposium on body composition convened by Brozek (1963) for the New York Academy of Sciences is an excellent guide to some of the more elegant procedures that are currently in use.

The suitability of any mathematical procedure must obviously depend upon the extent to which it describes the actual course of events in a meaningful biological sense. The cyclical growth curves generated by the ungulates we have studied present very real problems when one attempts to give them expression in equation form. This, coupled with the fact that a multitude of curve forms can be created by manipulating the nutritional environment, has led us to conclude that, until more data becomes available, little purpose is served by the application of elaborate mathematical procedures to the growth curves of the wild herbivores. On the basis of our experience to date, it would seem that a simple course of growth curve provides the most meaningful expression of events. There may be some merit in expressing growth rate in terms of time or feed intake over comparable phases of the cyclical growth curve when inter and intra species comparisons are considered desirable. In such cases simple arithmetic means derived from measurements on many individuals reared under comparable conditions are probably more meaningful than more elaborate mathematical expressions.

In the case of linear measures of growth, such as skeletal size, it is likely that logarithmic regressions will be more meaningful than simple linear regression. The heterogonic nature of skeletal growth supports the use of mathematical procedures that will permit the introduction of relative rate concepts into the expression of growth. While it is self-evident, it may be wise to point out that body weight can respond to environmental changes in both positive and negative directions whereas the linear measurements dependent on skeletal size can only be equal to or greater than the original measurement.

### References

BANDY P.J. (1965). A study of comparative growth in four races of black tailed deer. Doctorate Dissertation, The University of British Columbia, Vancouver, B.C., Canada. 189 p.

BANDY P.J., COWAN I.McT. & WOOD A.J. (1956). A method for the assessment of the nutritional status of wild ungulates. *Can. J. Zool.* **34,** 48–52.

BRODY, SAMUEL. (1945). *Bioenergetics and Growth.* Reinhold, New York. Republished 1964, Hoener Publishing Co., New York.

BROZEK, JOSEPH. (1963). Body composition. *Annals of the New York Academy of Sciences,* **110,** Parts I & II, 1–1018.

COWAN I.McT. & LOW W. (1963). Age determination of deer by annular structure of dental cementum. *J. Wildl. Mgt.* **27,** 466–471.

HAMMOND, JOHN. (1954). *Progress in the Physiology of Farm Animals.* Vol I and II, Butterworth, London.

HUXLEY J.S. (1932). *Problems of Relative Growth.* Lincoln MacKeogh, New York, The Dial Press.

SIR D'ARCY THOMPSON. (1942). *On Growth and Form.* The University Press, Cambridge, U.K.

SIR D'ARCY THOMPSON. (1961). *On Growth and Form.* Abridged Edition Edited by J.T. Bonner. Cambridge University Press, Cambridge, U.K.

WILSON P.N. & OSBOURN D.F. (1960). Compensatory growth after under-nutrition in Mammals and Birds. *Biol. Rev.* **35,** 324–363.

WINCHESTER C.F. (1964). Symposium on growth, environment and growth. *J. Anim. Sci.* **23,** 254–264.

WOOD A.J. (1964). Early weaning and growth of the pig. *Nutrition: Proc. Sixth Internat. Cong. Edinburgh,* edited by C.F. Mills and R. Passmore, (1963). E. & S. Livingston Ltd, Edinburgh.

WOOD A.J., NORDAN H.C. & COWAN I. McT. (1962). Periodicity of growth in ungulates as shown by deer of the genus *Odocoileus. Can. J. Zool.* **40,** 593–603.

# Section IV

# Management for Secondary Production

# 8

# The Husbandry of Wild Animals

## A M Harthoorn

Where man has domesticated animals he has undertaken to provide an environment most suitable for their reproduction and production. With advantages of concentrated feeding, shelter, and medicaments, cattle are capable of remarkable production. Where we cannot alter the environment in any significant way, the animals that are accustomed to the environment may be the most suitable animals to use the land.

This is particularly pertinent where disease is a major limiting factor to the introduction of domestic stock. It may appear that only one major type of disease need be eliminated before the country can be made suitable for cattle. The prevalence of a major enzootic may be misleading. In Texas *anaplasmosis*, causing a widespread loss in bovines was not recognized until *Piroplasmosis* was completely eradicated by the elimination of the Texas Fever Tick (Rhoad, 1955).

In virtually limitless areas such as many of the vast tracts in Africa in which *Trypanosomiasis* is enzootic, simple cropping operations may be carried out with few requirements in the way of management. If the area is circumscribed and the natural movements of the animals hampered or prevented, the conditions in such an area are no longer natural. When the natural conditions are disturbed, we must see that there is a minimum of harmful effect on the animals we are attempting to use. We should seek to balance the adverse effect of our interference as well as that of the harvesting itself by limiting the impact of the natural deleterious influences that normally keep the numbers in check.

Under management conditions a number of factors may operate to the detriment of the animal populations. Relative isolation and consequent inbreeding may induce heredity defects and lethals; restriction of movement to favored areas may induce mineral deficiencies; reduction of the number of species will increase the relative density of the favored animals and risk of disease; reduction of males by killing will increase the risk of breeding diseases; shooting of the best males in the herds is likely to induce degeneration.

Under these conditions the simple precepts of *management* which basically seek to maintain the balance between the animal species or between animals and vegetation may not be sufficient to ensure healthy stock and maximum productivity. The active control, for instance of breeding diseases by the exercise of veterinary principles of diagnosis and elimination, is a more specialized activity and more akin to the methods employed in husbandry of domestic stock. For this reason, the term *husbandry* has been used here to describe these aspects on routine ranching and medicine that may under certain conditions be applied advantageously to wild stock. These can be little more than listed in the space available, but reference works are cited in the bibliography below.

**1. Disease control.** Little is known of the diseases that affect wild animals. Proper slaughter of carcasses will yield much information that is both of essential scientific interest and a guide to management policies. When animals are restricted in their movements, the species variety reduced, numbers increased through protection, and where there is contact with domestic stock, manifestation of disease should be anticipated.

The epizootic diseases are of prime importance and lesions of infectious diseases, such as *foot-and-mouth* disease, should be watched for. Breeding diseases may occur partly as a result of management practices (*see* '2' below). Deficiency diseases are also important.

The problem is complex. As a general guide all untoward deaths should be investigated. Carcasses should be carefully inspected by competent personnel. Diseased or barren animals should be culled and subjected to a detailed examination immediately after death.

All ranching schemes should employ at least one person who has been trained, or is thoroughly conversant with the technique of carcass inspection and the principles of collection of specimens for laboratory diagnosis.

Experience in other countries dictates that an undiagnosed disease which is allowed to get a firm hold on a community may be extremely difficult and expensive if not impossible to eradicate.

The basic principles of hygiene must be observed in the slaughter of all meat for human consumption. It should be remembered that game farming has its opponents, and an outbreak of a disease in a city attributed to game meat may constitute a severe set back to game utilization.

**2. Management in relation to breeding.** One of the essentials of good stock is that it breeds. Some culling procedures are orientated towards eliminating non-breeders, poor stock, or diseased animals. A casual glance before culling may be misleading. A female may look well because she is barren and has not fed a calf, while another may show the drain of feeding one or several offspring.

Where numbers are to be reduced, one may wish to kill animals which will provide best quality meat. The judging of good condition comes with practice. Certain areas of the animal are diagnostic depending on the species. Guides have been published (Riney, 1960). There is the danger of one persistently killing the best breeding stock, and 'selecting' the breeders for unthriftiness. The tendency to do this is accentuated by the habit of good bulls to remain behind while the herd moves off. Therefore, the bulls of a herd should not be shot indiscriminately.

Where the males are excessively reduced, a certain amount of inbreeding will result, particularly if the herds are isolated. This *may* result in a deterioration or the manifestation of recessives leading to conditions such as sterility. In many instances, it will cause no appreciable harm. It should be remembered, however, that a tendency to homozygosity reduces the powers of adaptation to changes in the environment.

*See also* sections 1, 3, and 7 for factors that may influence breeding.

**3. Feeding during drought.** Cattle are fed during drought, and deer, in temperate countries, during the winter. The feeding of animals under severe drought conditions is therefore not beyond possibility, especially to preserve valuable breeding stock. During drought in Africa, however, fodder is scarce and of poor quality. Careful husbandry in the form of areas of grass or browse in reserve may constitute the best way of combating famine conditions. Browse is of particular importance as leaves may retain their nutritional value at a moderately high level long into the dry season after the grasses have dried out (Williamson and Payne, 1959.)

When congregation of stock in one area is induced through artificial feeding, possible increase in parasite burdens should be looked for or prevented by using fresh ground.

It is largely during reproduction and lactation that stress such as a drought takes a heavy toll. A whole calf crop may be lost during drought conditions.

**4. Provision of water.** There are a number of ways for providing water in arid areas. This is a highly specialized subject and cannot be detailed here. A point of particular interest is, however, the use of watering points to attract game to an area and to concentrate animals for counting, observing, marking, capture, or dosing. When watering points are closed, they may be used to drive game out of an area to rest the vegetation or to provide a reserve against shortage; and, whenever possible, watering places should be constructed or sited to render this possible.

It should be remembered that watering places around which game congregate may become focal points of disease conditions. Therefore, they should not be constructed carelessly or indiscriminately. Also, they may attract domestic cattle into areas which they were unable to use previously.

Irrigation is impractical for large areas, but it should be remembered that it was used to stimulate bush growth to feed rhinoceros in the Tsavo National Park during the 1961 drought.

**5. Fencing and corraling.** The herding of wild animals by means of fences has been hardly attempted in Africa. According to Beattie (1961) semi-wild cattle in Australia can be trapped at watering points in a corral with a funnel-shaped entrance. Since zebra can be driven by horses into corrals for capture, it is possible that other animals might respect corral walls. A modified funnel system can be used to get animals from one side of a fence to another either by driving or attracting through providing water. Effective vermin-proof fences have been constructed at Timau on the farm of Mr. Anthony Dyer with sheepwire put about 9 in. into the ground with about 3 ft. above ground which is fixed against five strands of high tensile wire fastened to 7 ft. posts which are buried two feet in the ground. Three foot long *podo* beams are fixed at 40° angle off the top of the posts to hold 8 strands of barbed wire. This wire leans outward to face the side from which game comes. This fence has been a paying proposition to keep out lion and leopard. It also successfully keeps out all other game including elephant. Prompt repair of breaks is absolutely essential.

Far less elaborate fencing may be effective if patroled and flanked by a road. All fences should be maintained until game trails are altered to conform with the obstacle. Ditches, about 6 ft. deep and 10 ft. wide, may be effective against some animals such as elephant. A single strand of wire should be placed on light posts on the floor of the ditch, *see* May (1964).

**6. Gauging of offtake on populations.** The number of animals shot in any breeding season should be consistent with the maintenance of numbers or increase or decrease as required.

Many factors influence the dynamics of a population, and the accurate assessment of offtake is a highly specialized procedure.

Careful cropping with repeated counts will preclude or minimize errors.

**7. Deficiencies and their diagnosis.** There are many places in the world where the supplying of minute amounts of trace elements, particularly cobalt, makes a difference between profit and loss in ranching (Rhoad, 1955). Without doubt the same applies to game ranching. Wild animals have been shown to suffer from deficiencies (Jarrett *et al.*, 1964) and a lack of trace elements should be guarded against.

Lack of phosphates or iodine can cause drastic curtailment of reproductive rates in domestic cattle.

The deficiencies may be peculiar for one species in the area, they may be seasonally associated with a flush of vegetation, or they may be induced by other factors such as a lack of condition or parasite burdens.

The diagnosis of low grade mineral deficiencies is often difficult. It may be reflected by a failure to reach maximum productivity. Most animals come readily to salt licks; and deficiencies of phosphorus, for instance, can be remedied either by the provision of sterilized bone meal or sodium phosphate.

**8. Improvement of range: burning.** In certain seasons when rain stops suddenly before the grass has matured and hot dry winds arrive, the grass growth may be rapidly arrested and yet the leaves retain their value. This is termed self-haying by grazers in Australia (Beatty, 1954). The burning of such stands is destruction of valuable fodder. Where grass grows to maturity, however, it may lose its feeding value; although, grass which has reached maturity will seed and enable re-seeding to take place.

The control of burning will usually provide control of bush. A portion of the land in bush is a safeguard against famine after the grass has gone in the dry season. Domestic stock are able to fall back on trees and shrubs when the grass is eaten, and presumably many wild animals that prefer to graze can also use these plants. Leaves and seeds from certain bushes and trees provide valuable fodder and possibly contain trace elements that are scarce in grass. Leaves and seeds may equally well be eaten after they have fallen to the ground.

**9. Driving.** Large scale attempts to drive and corral wild animals to remove them from certain areas (such as the one in Nabashose in Uganda) have failed, but some trappers manage to capture limited numbers of animals in this way with success. The failure of drivers may be due to two principal factors: first, a great deal of knowledge is needed to site the corrals, and this has never been compiled in Africa; and second, stockades must be at least old and weathered enough to lose the trace of human scent and disturbance. At the time of this writing (1965/6) a great deal is being learned in attempts to drive animals such as Thompson gazelle, eland and hartebeest, on a scientific project near Nairobi (D. Hopcraft); and the facts will provide a basis of essential recorded information.

Driving of animals such as bison in the United States and zebra in Africa is practicable. It seems a logical deduction that this can be done with the other large African animals when we discover the proper methods.

**10. Blood sampling and vaccination.** It is questionable whether ranching or use of wild animals on a semi-intensive basis will ever be successful without the knowledge of how to handle the animals at specified times. This will only be known after a great deal of research. Vaccination is one of the foremost reasons for handling wild stock. Until the threat of epidemics such as *rinderpest* is eliminated, the ranching of wild animals will remain a danger and financial hazard.

It seems likely that driving of wild animals through chutes will become feasible before vaccines against diseases such as *rinderpest* will be developed for oral use. Application by aerosols is equally unlikely to be practicable and infection by projectile too time consuming. Furthermore, the ability to drive and restrain in corral, chute or crush will enable other operations, such as blood sampling, to be carried out. This will be of benefit in the control of diseases such as *contageous abortion* that may be readily transferred from cattle to certain and possibly many wild animals. Also, control of other diseases such as *malignant catarrh* in young wildebeest seems remote unless vaccine can be administered to groups of animals to act as immune barriers.

**11. Range management.** Areas may be sealed off from grazing by closing watering points. This may be done to prevent over-grazing or to allow grass to reach maturity for re-seeding. Re-seeding also occurs where grasses mature under the branches of fallen trees and are not able to be reached for grazing.

Grasses growing in tufts can sometimes be controlled under domestic conditions by grazing with sheep. Increased density of certain small wild grazers may have similar effect. In very limited areas a close sward can be induced by cutting the grass. This attracts impala and warthog which tend to keep the grass short. This may be used as limited management of the pasture or to attract animals to a confined area.

Management policies should determine the species of animals to provide the best crop in various areas. Local conditions will influence the choice.

Licks may increase and become troublesome as a result of the restriction of burning.

**12. Herd management.** Herd management centers largely around breeding and nutrition. The optimum proportion of bulls to cows in the various species will need to be determined. The disturbance of cropping should be timed to exclude the calfing period. Abandoning of calves under domestic conditions is sometimes due to nutritional factors. Inspection of carcasses (Ledger, 1964) will give indication of adequate nutrition. The percentage of females bearing young is an important indication of good or bad husbandry, as is the number of offspring surviving.

It is important to cull barren females to check for breeding diseases such as *contagious abortion*. Other breeding diseases such as *vibrio foetus* and *trichomoniasis* have not been reported as affecting wild animals.

Regular examination will give indication and warning of an increase in parasite burdens. Some animals are more prone to internal parasitism and tick infestation. A build-up of these may be discouraged by reducing the number of susceptible animals in relation to others. Some species have a 'cleaning' effect on the internal parasites of others.

### References

BEATTIE W.A. (1954). *Beef Cattle and Management.* The Pastoral Review Pty. Ltd., Melbourne.

JARRETT W.H.F., JENNINGS F.W., MURRAY M. & HARTHOORN A.M. (1964). Muscular systrophy in wild Hunter's Antelope. *E. Afr. Wildl. J.* **2**, 158–159.

LEDGER H.P. (1964). The role of wildlife in African agriculture. *First FAO African Regional Meeting on Animal Production and Health*, Addis Ababa, Ethiopia. March, 1964.

LEES M.T. (1964). Movement control of wildlife with reference to game fences. *First AFO African Regional Meeting on Animal Production and Health*, Addis Ababa, Ethiopia. March, 1964.

RHOAD A.O. (1955). *Breeding Beef Cattle for Unfavorable Environments.* A Symposium presented at the King Ranch Centennial Conference. Univ. of Texas Press.

RINEY T. (1964). The economic use of wildlife in terms of its productivity and its development as an agricultural activity. *First FAO African Regional Meeting on Animal Production and Health*, Addis Ababa, Ethiopia. March, 1964.

WILLIAMSON G. & PAYNE W.J.A. (1959). *Animal Husbandry in the Tropics*. Longmans, Green & Co., Ltd., London.

# 9

# Handling Wild Herbivores

## PART I. MANIPULATION OF ANIMALS FOR EXPERIMENTAL PURPOSES

### A M HARTHOORN

### Manual Restraint and Training

The restraint of undomesticated herbivores may be broadly divided into methods for captive and free animals.

The degree of restraint needed for captive animals differs according to the extent to which they have been tamed. If sufficiently tractable, the methods commonly used for domestic stock—such as a crush— may be useful. They are described fully in standard text books (e.g. Leaky and Barrow, 1953).

Captured eland (*Taurotragus oryx*), hartebeest (*Alcelaphus busephalus*), Wildebeest (*Connochaetes taurinus*) and buffalo (*Syncerus caffer*) have been trained to enter a crush with a 'V' shaped entrance. Provided the crush is sufficiently high and well built, the animals stand quite still for fairly extensive manipulations, including rectal temperature measurements and blood sampling. Once trained, wild animals may enter a crush with much less protest than cattle, but subject to certain precautions. The animals should be driven into the crush as far as possible by the same attendant. They will usually stand quiet only when in a chosen order. No animal should jump out of the crush, or the training programme receives a set back. The routine must be continuous. Even though quiet and ostensibly willing, mature antelope are always dangerous and may inflict injury without harmful intent.

Even untrained herbivores will readily come to certain sections of the enclosure for food; and surprisingly extensive manipulations may be carried out while the animals are engrossed with eating—such as rectal exploration, and blood sampling of captive rhinoceros.

Chemical restraint may be necessary. When this may interfere with the validity of the samples, catheters may be placed in arteries or veins, into the heart and pulmonary artery; and blood samples taken for several weeks or longer. The exterior part of the catheter may be led under the skin for long distances such as from the carotid artery to the shoulder or back, for ease of sampling. Data such as blood pressure, deep body temperature, heart rate and respiration are conveniently obtained from the undisturbed animals with the use of radiotelemetric devices. These can be expected to transmit valid physiological data about 24 hours after placement under the effect of re-straining drugs and continue to transmit for variable periods according to the battery life. Completely wild animals may be monitored in a similar way. An advantageously placed long-distance sender may be added to the instru-mentation for location prior to monitoring, or else a receiving-transmitting device that picks up signals from other apparatus for retransmission. Wild animals, even oryx (*Oryx beisa*) and ostrich (*Struthio camelus*), become remarkedly tolerant of apparatus strapped or glued onto the back. They can also be trained to accept being tied to breath analysing apparatus or connec-tions to overhead swivels, although under these circumstances the reactions of the animals may differ from the normal.

For newly captured animals the shape of the enclosure is all important and freshly captured hoofed animals may be released into the 10 foot wide space between two concentric rings of (say) 8 foot high reed matting (with diameters of 15 and 25 feet). By trying to break through or jump the walls, they may suffer heavy mortality if placed in a simple rectangular enclosure.

## Chemical restraint

The chemical names and sources of the drugs mentioned below by their trade names are listed at the end of this section.

**Captive animals.** This must be dealt with briefly, as the degree of restraint in relation to training offers infinite variations. Where animals are completely intractable, and in grazing paddocks, they may have to be treated as free animals (*see* below) although tranquilizers or sedatives may be given to captive animals over a period of days or weeks and preferably incorporated in food or water.

One may administer the following according to the temperament of the animal and to the extent of the interference:

Psychomotor drugs, e.g. *Diazapam*. These often excite, but depress in large doses. They may appear to tranquillize but the animal may become violent when forcibly restrained. These compounds are most useful to accustom animals to a new routine and surroundings and for minor manipulations.

The compound *Quiloflex* has a central sedative action with a marked inhibitory effect on muscle reflexes. Good results have been obtained on captive wild animals being handled for various purposes in doses of 1·0 mg/kg upwards. At higher doses of several mg/kg there is a tendency to bloat, but as the righting reflexes are retained there is little danger. Numbers of antelope and gazelle have been manipulated with *Quiloflex*, sometimes with the addition of *Largactil* or *Acepromazine*. It potentiates the tranquilizing effect of these substances. *Quiloflex* has also been used successfully on buffalo and zebra (*Equus burchelli*) and on giraffe (*Giraffa camelopardalis*). At 2 mg/kg animals have been strapped to an operating table and superficial operations performed under local anaesthesia for the placement of radio-telemetric sensors. It has been used for the capture of wild animals at doses of 10—20 mg/kg by van Niekerk *et al.* (1963). An operating table built on the principle of a railway man's trolley and pulled as a trailer has been found invaluable. The animals are strapped against it and the table tipped to the horizontal position once the standing animal is secured.

Tranquilizers; the main indication is to use concurrently with synthetic morphine-like compounds used for immobilisation.

Major tranquilizers are substances like the butyrophenone derivatives such as *Fluanisone, Aceperone, Droperidol* and *Haloperidol* and also *Phencyclidine*.

Minor tranquilizer; e.g. chlorpromazine, acetylpromazine, trifluopromazine. These produce a potentiating effect on the morphine-like substances such as *Etorphine* (M.99) or *Fentanyl*. For some animals such as nyala (*Tragalaphus ngasi*), kudu (*Tragalaphus strepsiceros*), and others major tranquilizers should be resorted to. Fluanisone and related substances cause less disruption of the heat regulating mechanism than do the phenothiazines. (Pienaar *et al.*, 1966.)

Some ataractic compounds; e.g. chlorpromazine. Drowsiness is induced by phenothiazines without a piperazine chain (Freyhan, 1959), such as chlorpromazine, trifluopromazine, acetylpromazine etc. These combat excitement and have a sedative effect. There is increased heat exchange due to peripheral vasodilatation, and temperature control is also impaired.

Sedatives; e.g. Chloral hydrate. It is not easy to induce animals to take this by mouth as it is caustic. It is useful to stop bad habits induced by restlessness. One ounce a day given in brewers' residues works excellently to prevent habitual stall breaking by a bull buffalo b.w. 600 kg. In large doses it will cause anaesthesia or loss of balance and falling. It is cheap and easily procured.

**Wild animals.** Succinylcholine chloride and related compounds have been used successfully on a large number of species. It has a small safety margin and its use in the field is often attended by a high mortality rate, except in certain species, or when the operator restricts himself to one age group in a species. Body weight can be estimated sufficiently accurately only when many of one type of one animal are required. Succinylcholine is thus unsuitable for use sporadically on large wild mammals. Antidotes for succinylcholine such as *Cholase* (Evans *et al.*, 1953) and certain experimental quaternary compounds are not generally useful. If *effective* artificial respiration can be given, mortality should be very low. A small oxygen cylinder, with a needle for tracheal puncture, or endotracheal tube, may be advantageously carried to counteract anoxia from respiratory paralysis. Cardiac irregularities and deaths due to heart failure as a result of succinylcholine injection have been reported in the domestic horse. Artificial respiration without mechanical aids is seldom effective in animals weighing over 100 kg. due to the very large dead space in the respiratory tract.

In man the effect of succinylcholine injection into the conscious subject is frightening and painful. Death from overdosage is caused by asphyxia due to respiratory paralysis, and it is inhumane to use a neuromuscular blocking agent for euthanasia or cropping.

*Flaxedil* is more suitable than succinylcholine for use on the larger animals such as giraffe, being more sparing of the respiration and having a wider safety margin. Even so, the difficulties and dangers attendant on the use of the antidote *Prostigmin* and the chances of subsequent relapse make this compound unsuitable for general use on wild animals. Ruminants tend to regurgitate.

For acute restraint of captive animals *Flaxedil* is efficacious in doses designed only to weaken. Animals are unable to resist after intramuscular injection at about 1·0 mg/kg body weight. This has been used successfully to implant sensors under local anaesthesia in mature and recalcitrant buffalo.

*Nicotine* has very few indications for animal capture and may cause intense distress and high mortality. Nicotine alkaloids are extremely poisonous and therefore dangerous to the operator.

**Centrally acting immobilising drugs.** With the use of tranquilizer/morphine/ scopolamine mixtures losses from non-recovery in large African mammals were reduced virtually to zero. Well over a hundred white (square-lipped) rhinoceros (*Ceratotherium simum*) were relocated using a tranquilizer/thiambutene/scopolamine combination without a fatality due to drug effects (Harthoorn and Player, 1963, and Harthoorn, 1967).

The principal disadvantage of Themalon is its large bulk and consequent slow absorption. It has been replaced since 1963 with the M. series of compounds such as Etorphine (M.99) particularly for wild animals.

As with all morphine-like substances, Etorphine causes a degree of excitement and, in animals such as the horse, tachycardia and a rise in blood pressure. Excitement may be largely controlled by tranquilizers and adrenolytics. The amount of tranquilizer that need be added differs among the species from none or a very little for elephant, rhinoceros, and hippopotamus, about 40 mg per animal of acepromazine for zebra, wildebeest, etc., to a large quantity for antelope such as nyala and oryx, and eland. Details of dosages can be found in articles by Pienaar *et al.* (1966); Harthoorn and Bligh (1965); Player, 1967; and King and Klingel, 1966.

The dose given becomes more effective if dissolved in dimethylsulphoxide (DMSO), especially for elephant, but the danger of inadvertent absorption through the human skin becomes much greater than if dissolved in water.

Fentanyl is a morphine-like substance that is gaining popularity as an immobilising compound. It is somewhat less than 1/10th the strength of Etorphine. Fentanyl has certain disadvantages for field work in that it is relatively insoluble, but solutions containing 40 mg/ml of Fentanyl citrate may be dissolved in DMSO (Pienaar, 1967).

With the use of Etorphine (as with morphine, *Themalon* and Fentanyl) captured animals may be handled while standing. If down, they may be brought to their feet by a graduated dose of *Lethidrone* or by the more specific antagonist *Cyprenorphine* (M.285). Wild elephants will stand through experimental procedures that involve mounting with a ladder; zebra and antelope have been led for long distances to a weighing platform, and giraffe to holding enclosures. The use of M.183 rather than Etorphine (M.99) has been advocated for giraffe (Pinaar *et al.*, 1966).

Etorphine may be used in a solution of 5 mg/ml in water acidified to pH 4·5 with hydrochloric acid, or 10 mg/ml in DMSO. In water, solubility is decreased to only 1·2 mg/ml at 20° by addition of scopolamine solution at 100 mg/ml due to common ion effect. The use of scopolamine with Etorphine has, however, largely been discontinued.

**Drugs**

*Acepromazine.* Acepromazine maleate
    Boots Pure Drug Co Ltd, Nottingham, England

*Azaperone.* Flupyridol. 1-(3-4-fluoro-(bensoyl)-propyl)-4-(2-pyridil)-piper-
azine
    Janssen Pharmaceutica, Turnhout, Belgium

*Cholase.* Pseudo-cholinesterase
    Cutter Laboratories, Berkley, California, U.S.A.

*Diazapam.* Benzodiasepine derivative RO 5–2807
    Hoffmann-La Roche, Basle, Switzerland

*Fentanyl.* 1-(2-phenethyl)-4-(N-propionyl-anilino)-piperidine
    Janssen Pharmaceutica, Turnhout, Belgium

*Flaxedil.* Gallamine triethiodide
    May & Baker Ltd, Dagenham, England

*Droperidol.* Dehydrobenzperidol. 1-(3-(4-fluoro-(benzoyl)-propyl)-4-(2-
oxo-1-benzimidazoli-nyl)-1,2,3,6-tetrahydropyridine.
    Janssen Pharmaceutica

*Fluanisone.* 1-(3-(4-fluoro-benzoyl)-propyl)-4-(2-methoxy-phenyl)-piper-
azine
    Janssen Pharmaceutica

*Haloperidol.* 'Serenance' 4-fluoro-4-(4-hydroxy-4(4-chlorophenyl) piper-
idino butyrophenone
    G.D. Searle & Co., Chicago, Illinois, U.S.A.

*Largactil.* Chlorpromazine hydrochloride
    May & Baker Ltd, Dagenham, England

*Lethidrone.* Nalorphine (N-allynormorphine)
    Burroughs Wellcome & Co, London, England

*Prostigmin.* Neostigmine methylsulphate
    Roche Products Ltd, Welwyn Garden City, Herts, England

*Quiloflex.* Benzodioxane hydrochloride
    Boehringer u Soehne, Mannheim, Germany

*Themalon.* Diethylthiambutene
    Burroughs Wellcome & Co, London, England

*Etorphine.* M.99-(6:14 *endo*etheno-7-$\alpha$(-2-hydroxy-2-pentyl)-tetrahydro-
oripavine hydrochloride*
    Reckitt & Sons Ltd, Hull, England

*Slightly different formulae have been issued by the British Pharmacopoea Commission
and in the United States.

*Cyprenorphine.* M.285 N-cyclopropylmethyl-6:4-*endo*etheno-7-(2-hydroxy-2-propyl)-tetrahydro-nororipavine hydrochloride
Reckitt & Sons Ltd, Hull, England

### References

EVANS F.T., GRAY P.W.S., LEHMANN H. & SILK E. (1953). Effect of pseudo-cholinesterase level on action of succinylcholine in man. *Br. Med. J.*, **1**, 136–138.

FREYHAN F.A. (1959). Therapeutic implications of differential effects of new phenothiazine compounds. *Am. J. Psychiat.* **115**, 577–585.

HARTHOORN A.M. (1967). Comparative pharmacological reactions of certain wild and domestic mammals to thebaine derivatives in the M. series of compounds. *Fed. Proc.* **26**, 1215–1261.

HARTHOORN A.M. & BLIGH J. (1965). The use of a new oripavine derivative with potent morphine-like activity for the restraint of hoofed wild animals. *Res. Vet. Sci.* **6**, 290–299.

HARTHOORN A.M. & PLAYER I.C. (1963). The narcosis of the white rhinoceros. A series of eighteen case histories. *Proc. Zoological Congress, Amsterdam.*

KING J.M. & KLINGEL H. (1966). The use of the oripavine derivative, M.99, for the restraint of equine animals, and its antagonism with the related compound M.285. *Re.. Vet. Sci.* **6**, 447–455.

LEAKY J.R. & BARROW P. (1953) *Restraint of Animals.* 2nd ed. Cornell Campus Store, Inc., Ithaca, N.Y.

PIENAAR U. DE V. (1966). Capture and immobilizing techniques currently employed in Kruger National Park and other South African national parks and provincial reserves. Kruger Nat. Park, P.O. Skukuza, South Africa. 9 p. mineo.

PIENAAR U. DE V. (1967). Recent advances in the field immobilization and restraint of wild ungulates in South African National Parks. Kruger Nat. Park, P.O. Skukuza, South Africa. 13 p. mimeo.

PIENAAR U. DE V., VAN NIEKERK J.W., YOUNG E. & VAN WYK P. (1966). Neuroleptic narcosis of large wild herbivores in South African National Parks with the new potent morphine analogues M.99 and M.183. *J. S. Afr. Vet. Med. Assoc.* **37**, 277–291.

PIENAAR U. DE V., VAN NIEKERK J.W., YOUNG E., VAN WYK P. & FAIRALL N. (1966). The use of oripavine hydrochloride (M.99) in the drug-immobilization and marking of wild elephant (*Loxodonta africana Blumenbach*) in the Kruger National Park. *J. Sci. Res. Nat. Parks S. Afr.* (*Koedoe*). **9**, 108–124.

PLAYER I. (1967). Translocation of white rhinoceros in South Africa. *Oryx, IX.* **2**, 137–150.

VAN NIEKERK J.W., PIENAAR U. DE V. & FAIRALL N. (1963). A preliminary report on the use of Quiloflex (Benzodioxane hydrochloride) in the immobilization of game. *J. Sci. Res. Nat. Pks. S. Afr.* (*Koedoe*). **6**, 109.

See also:

HARTHOORN A.M. (1965). The application of pharmacological and physiological principles in restraint of wild animals. *Wildl. Monog. No.* 14. The Wildlife Society, Wisconsin Ave., Washington, D.C.

F.A.O. Monograph on the *Animal Health Series*. (In preparation.)

## PART II. CAPTURE AND IMMOBILIZING TECHNIQUES CURRENTLY EMPLOYED IN SOUTH AFRICAN NATIONAL PARKS AND RESERVES

U de V Pienaar

1. Small antelope, including steenbuck (*Raphicerus campestris capricornis* (Thomas and Schwann)), Sharpe's steenbuck (*Raphicerus melanotis sharpei* (Thomas)), Red duiker (*Cephalophus natalensis amoenus* (Wroughton)), Grey duiker (*Sylvicapra grimmea caffra* (Fitzinger)), Oribi (*Ourebia ourebi ourebi* (Zimmerman)), Klipspringer (*Oreotragus oreotragus transvaalensis* (Roberts)), Mountain reedbuck (*Redunca fulvorufula fulvorufula* (Afzelius)), Grey rhebuck (*Pelea capreolus* (Forster)), and even reedbuck (*Redunca arundinum arundinum* (Boddaert)), are best captured by driving them into nets in the manner described by Pienaar and Van Niekerk (1963). This is best accomplished with the aid of numerous beaters and preferably also, where the terrain is suitable, a number of men on horseback.

Injuries are minimal, particularly if an experienced catching team is employed, and losses in transit are rare.

Captured animals are immediately blindfolded with a strip of black cloth and the feet tied temporarily with nylon stockings. A tranquilizing drug is administered at this stage, and a long-acting penicillin such as 'Triplopen' (Glaxo-Allenbury's) is routinely also injected intra-muscularly. Glucocorticosteroids may be administered, but it is usually not necessary.

When the animals have calmed down sufficiently, they are laid down gently in a sternal position on open grain bags, which are then folded over the back of the animal and stitched up with a thatching needle. At the same time their feet are released. The animals soon shuffle into a comfortable position, but are prevented by their grainbag-jackets to stand up or thrash about.

The trussed-up animals are then removed and placed on a deep bed of straw on the back of the transporting vehicle.

In this manner, captive animals have been held for periods of up to 12 hours and transported for several hundred miles. For longer journeys it may be advisable to place each captive animal in suitably dark and ventilated crates of correct size and construction. These crates should have sliding doors at both ends to facilitate easy release.

Various tranquilizing drugs have been used for the restraint of captive small antelope, and of these chlorpromazine (Largactil) at a dosage rate of

0·75—1·0 mgm/lb. has been widely advocated. The most universally satisfactory results have been obtained with trifluomeprazine ('Nortran'—Norden Laboratories Inc., Lincoln, Nebraska, U.S.A.), at a dosage rate of 0·5—1·0 mgm/lb.

Before release, it is advisable to dose animals orally with a safe anti-helmintic such as thiabendazole; and if the quality and composition of the vegetatio n of the new habitat varies much from that of the area of capture, it is essential to dose the animals orally with a few ccs. of rumenal fluid of some local ruminant or the fresh dung of local ruminants shaken up in water, in order to prevent fatal digestive disturbances.

**2.** Medium-sized plains-loving ungulates, including springbuck (*Antidorcas marsupialis marsupialis* (Zimmermann)), blesbok (*Damaliscus dorcas phillipsi* (Harper)), bontbok (*Damaliscus dorcas dorcas* (Pallas)), red hartebeest (*Alcephalus buselaphus caama* (Cuvier)), and black (or white-tailed) wildebeest (*Connochaetes gnou* (Zimmermann)), may be captured in large numbers by driving them into holding pens and crushes with the aid of vehicles and horsemen in the manner described by Riney and Kettlitz (1964). The technique is fraught with hazards of diverse nature and the losses through injury are inevitably considerable. Individual animals may be diverted from the main herd, run down by means of a suitable vehicle, and roped by means of a pole-lasso. This is a better technique, but more time-consuming. It is usually employed for the capture of the rare black wildebeest.

Captured animals are blindfolded and pushed into transporting crates which for all the above species, except springbuck, may be 24 inches wide, 5 feet high and 6½ feet in length, with sliding panels at both ends. For springbuck the crate size is 18 inches wide, 3 feet high and 4½ feet in length.

Tranquilizing drugs may or may not be of use and again chlorpromazine (Largactil) at 1 mgm/lb. has been universally employed with success. Even more satisfactory effects have been achieved by administering trifluopromazine ('Siquil'—Squibb) at a dosage rate varying from 0·05 to 0·1 mgm/lb.

Animals which have been chased over long distances should receive both a long-acting penicillin and glucocorticosteroid (e.g. 'Vecortenol' of Ciba).

Where conditions permit, these animals may also be captured by driving them into nets set up in a manner similar to that which is employed for the smaller antelope. The nets must obviously be of stouter construction and are usually made from braided nylon cord of ± 1000 lb. breaking strain.

Successes are often doubled if the net is not set in a straight line, but in such a manner that the various sections form an inverted J (Figure 9.1),

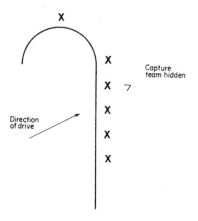

Figure 9.1.

or in the form of a battlement (Figure 9.2).

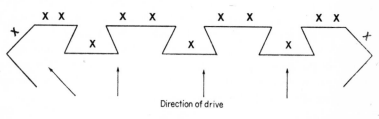

Direction of drive

Figure 9.2.

This latter setting of the net is particularly expedient for the capture of cunning forest dwellers which tend to double back during a drive. These include Nyala (*Tragelaphus* (*Nyala*) *angasi* (Gray)), bushbuck (*Tragelaphus scriptus roualeyni* (Gray)) and red duiker (*Cephalophus natalensis amoenus* (Wroughton)).

It may be possible in some small reserves or on farms, where selective capturing is desirable and where the animals are not particularly timid, to approach them within range of the various dart-guns or crossbows currently employed for drug-immobilization of animals.

In this event we have found the following drug combinations most successful for the plains game:

(i) RED HARTEBEEST:
ADULT MALES
1 mgm Etorphine (M–99) (Reckitt & Sons).
10—15 mgm Acetyl-promazine (Boots) (or 50—70 mgm. Azaperone (Janssen)).
5 mgm Hyoscine hydrobromide.

ADULT FEMALES
0·75 mgm M–99.
10 mgm Acetyl-promazine.
10 mgm Hyoscine hydrobromide.

(ii) BONTEBOK AND BLESBOK: ADULTS
1 mgm Etorphine (M–99).
10—15 mgm Acetyl-promazine (or 50—70 mgm Azaperone Janssen).
5 mgm Hyoscine hydrobromide.

Acetyl-promazine (Boots) is not a good tranquilizer for this species and may cause untoward reactions such as fatal interference with heat regulation. This is also the case with species with thick, shaggy coats such as waterbuck and reedbuck, as well as with other dark-skinned ungulates e.g. black and blue wildebeest, sable, and buffalo.

We are at present conducting field trials with new butyrophenone derivatives 'Fluanisone'* (Janssen) and 'Azaperone'** which appear to induce excellent tranquilization without the unwanted side reactions of Acetyl-promazine or the other phenothiazines. There is no interference with heat regulation; the potentiating effect on Etorphine is even greater than that of Acetyl-promazine and the tranquilizing effect, even post-Nalorphint administration, is amazing.

This is particularly true in the case of zebra, but very satisfactory results have also been obtained with blue wildebeest, roan antelope, impala and

*1-(3-(4-fluoro-benzoyl)-propyl)-4-(2-methoxy-phenyl)-piperazine (R 2028).
**1-(3-(4-fluoro-benzoyl)-propyl)-4-(2-pyridil) piperazine (R 1929).

buffalo. Presupposing a similar reaction in bontebok and blesbok, the recommended does of Fluanisone would be 10—15 mgm of Azaperone 50—70 mgm.

(iii) BLACK WILDEBEEST: ADULTS

1·5 mgm Etorphine (M–99).

15—20 mgm Fluanisone or 50 mgm Azaperone (alternatively 15—20 mgm Acetyl-promazine).

5—10 mgm Hyoscine hydrobromide.

(iv) SPRINGBUCK: ADULTS

0·25—0·5 Etorphine (M–99).

5 mgm Acetyl-promazine or 25 mgm Azaperone.

2·5 mgm Hyoscine hydrobromide.

The narcotic effect of the potent analgesic Etorphine (M–99) is satisfactorily and speedily antagonized in all these animals by Nalorphine hydrobromide (Lethidione of Burroughs Wellcome), administered intravenously (or, if not possible, intramuscularly).

A standard dose of antidote of 50 mgms. for small antelope and 100 mgms. (total dose) for medium-sized antelope was employed. Alternatively, the specific Etorphine antagonist, Cyprenorphine (M—285) Reckitts may be employed.

3. Impala (*Aepyceros melampus melampus* (Lichtenstein)) may be captured by hand in large numbers during moonless nights by blinding them with powerful sealed beam search lights and stalking them from the darkness by a darkly-clad catching team.

Impala captured in this manner are often transported in bulk in darkened and padded communal crates on the back of vehicles, but they should preferably also be crated individually as described above for springbuck. The use of tranquilizers is optional but administration of antibiotics can only cut down losses.

If it is desired to capture impala with the aid of drugs, the following combination may be used for adult animals:

0·5 mgm Etorphine (M–99), or, preferably, 10 mgm Fentanyl (Janssen)

5 mgm Acetyl-promazine maleate or, preferably, 25—30 mgm Azaperone (Janssen).

2·5 mgm Hyoscine hydrobromide.

Impala are rather sensitive to the action of Etorphine, and at this dosage rate they will usually lapse into a coma and die within 30 minutes if no remedial measures are applied. The intravenous infection of 10 mgms of Nalorphine hydrobromide immediately after capture, is normally sufficient to prevent any such occurrence.

In more open country the dose of Etorphine may well be cut to 0·25 mgm; and, although the animal will remain on its feet longer, there is no danger of excessive respiratory depression.

The addition of the enzyme Hyaluronidase (1500 i.u.) to the drug mixture usually facilitates a more rapid induction of the immobilizing effect.

Waterbuck (*Kobus ellipsiprymnus ellipsyprymnus* (Ogilby)) have been successfully captured with neuroleptic analgesic mixtures, but the therapeutic index of mixtures containing phenothiazines and hyoscine is not favorable in this species and losses are experienced from heat stroke and collapse, torticollis of the neck-musculature, cardiac failure, etc. Adult bulls have been successfully captured with 3—3·5 mgm Etorphine and adult cows with 2—2·5 mgm. Etorphine in combination with Acetyl-promazine 5 mgms or preferably Azaperone 100—150 mgms. Hyoscine causes toxic reactions in this species and should be omitted from drug mixtures.

Kudu (*Tragelaphus strepsiceros strepsiceros* (Pallas)) have been successfully immobilized with Gallamine triethiodid (Flaxedil) at dosage rates varying from 0·9—1·1 mgm/lb., but the safe range is rather critical and the risk of mortality great. Muscular paralysis may be reversed with Neostigmin (Prostigmin) at a level of 0·01 mgm/lb. The most satisfactory drug combination for kudu used by us to date is the following:

*Adult bulls:*

| | |
|---|---|
| Etorphine (M–99) | 4 mgm. |
| Sernylan | 75—100 mgm. |
| Trifluopromazine (Siquil) | 50 mgm. |

*Adult cows and young bulls:*

| | |
|---|---|
| Etorphine (M–99) | 2·5 to 3 mgm. |
| Sernylan | 50 mgm. |
| Trifluopromazine | 50 mgm. |

Hyoscine hydrobromide may also be added to this mixture at a dosage rate of 5 mgm/200 lb. body weight, but is best left out if the animal is to be released immediately after marking it for instance. This is done because hyoscine induces dilation of the pupil and the prolonged state of photophobia makes the animal easy prey for predators and handicaps its chances of survival.

Recent experiments indicate that the trifluopromazine may well find a more efficient substitute in Fluanisone or Azaperone (Janssen).

The same drug combination and dosage level employed for adult kudu bulls may also be used for eland cows (adults). For adult eland bulls (*Taurotragus oryx oryx* (Pallas)) the dose of Etorphine (M–99) must be increased to 5 mgm. and that of Azaperone to 200 or 300 mgm.

A provisional dose for adult sable antelope (*Hippotragus niger niger* (Harris)) in the light of limited experience would be: Etorphine 2 mgm, Hyoscine hydrobromide 5 mgm, Fluanisone 30—40 mgm (as alternative Azaperone 200 mgm). A similar drug combination is also proposed for roan antelope (*Hippotragus equinus equinus* (Desmarest)), but for adult bulls it may be necessary to increase the dose of Etrophine to 3 mgm or 4 mgm.

Dr. H. Ebedes of the Etosha Game Park (South West Africa) informs me that Gemsbuck (*Oryx gazella gazella* (Linnaeus)) are sometimes highly resistant to the action of Etorpine—particularly pregnant females. Adults have, however, been successfully captured with 3—4 mgm Etorphine in combination with Trifluopromazine 50 mgm or Azaperone 300 mgs and Hyoscine 10 mgs.

An effective immobilizing dose of Etorphine (M–99) for warthog (*Phacochoerus aethiopicus sundevalli* (Lonnberg)) is 1·0—1·5 mgm in combination with 5 mgm Hyoscine hydrobromide and 20 mgm Acetyl-promazine (or 50 mgm Azaperone).

Blue wildebeest (*Connochaetes* (*Gorgon*) *taurinus taurinus* (Burchell)) may be captured with almost infallible certainty and negligible loss with the following combination:

*Adults:*
Etorphine (M–99)           2 mgms.
Fluanisone                 20 mgms (alternative 20 mgms. Ac. promazine).
Hyoscine hydrobromide  20 mgms (optional).
1–2 *Year old young:*
Etorphine (M–99)           0·5–1 mgm.
Fluanisone                 10 mgms.
Hyoscine hydrobromide  10 mgms.

The same drug-mixtures and dosage levels may also be employed for the capture of zebra (*Equus* (*Hippotigris*) *burchelli antiquorum* (H. Smith)). In the case of these animals Fluanisone is definitely preferable to Acetyl-promazine or the phenothiazines and induces excellent tranquilization in the recovery phase post-nalorphine administration. When zebra have to be transported, their crates are best fitted with padded shoulder supports to prevent the animal from moving forward in characteristic manner during the recovery stage and injuring or breaking its neck (Figure 9.3).

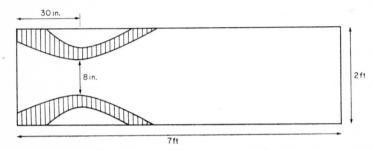

Figure 9.3. Crate for 1—2 year old zebra: height of crate 5 ft.

Tsessebe (*Damaliscus lunatus lunatus* (Burchell)) have been captured with 1 mgm. Etorphine (M–99), 10 mgm Hyoscine hydrobromide and 10—15 mgms. Acetylpromazine. It would be well if the latter were substituted with 10 mgms. Fluanisone, particularly on hot days, to obviate serious heat regulation disturbances.

An effective and safe immobilizing mixture for adult buffalo (*Syncerus caffer caffer* (Sparrman)), is the following:
*Adult bulls:*
4—5 mgms Etorphine (M–99).
80 mgms Fluanisone (20 mgms Acetyl-promazine).
10—20—100 mgms Hyoscine hydrobromide (optional).
*Adult females:*
3—4 mgm Etorphine (M–99).
60 mgms Fluanisone (or 200 mgm Azaperone).
10 mgms Hyoscine hydrobromide.
The usual dose of antagonist administered is 100—200 mgms of Nalorphine hydrobromide.

African elephants (*Loxodonta africana africana*, (Blumenbach)) are much more sensitive to the action of neuroleptic-analgesic mixtures than most other species. Compared to the dosage rate for Etorphine in the case of most ruminant species (2·0—4·0/u gm/lb.), that for elephant is very much lower (0·47—0·67/u gm/lb.).

The optimum dosage rate of Etorphine when combined with Acetyl-promazine would appear to be 7—8 mgms (total dose) in the case of the largest group of adult elephant bulls (weighing 12,000—15,000 lb.) and 5–6 mgms. for the smaller class adult bulls and largest cows (7,000—12,000 lb body weight). Etorphine solutions for immobilizing elephant are best made up in Dimethylsulphoxide instead of water acidified to pH 4 with NHCl. These solutions should be handled with great care as they may be absorbed through the skin.

Etorphine is administered in combination with Acetyl-promazine (50—60 mgms total dose for the largest bulls and 40—50 mgms for the smaller adult bulls). The latter may well be substituted by Fluanisone or Azaperone with even more satisfactory results.

Hyoscine hydrobromide apparently causes toxic reactions in elephant, as in the case of waterbuck, and it is best omitted from drug mixtures.

Large amounts of Nalorphine are necessary to antagonize the effect of Etorphine in these massive beasts and the antagonizing action is to some degree weight dependent. On the other hand, highly dependable reversal of narcosis is obtained in elephant with Cyprenorphine hydrochloride (M–285) and the optimum dose of antidote in the case of beasts immobilized with 5—8 mgms. Etorphine seems to be in the region of 40—60 mgms M–285. Elephants which go down in a sternal position when succumbing to the drug reaction should be pulled over on their sides by means of a rope and truck in the manner described by Pienaar et al. (1966) as soon as possible in order to prevent fatal respiratory and circulatory collapse.

The oripavine derivative Etorphine acetate (M–183) is a safer drug to use for the capture of giraffe (Giraffa camelopardalis giraffa (Boddaert)) than the related more potent Etorphine hydrochloride (M–99). The onset of its reaction is more gradual than that of M–99, and its effect is less drastic on cardiac function and respiration.

M–99 causes a drastic fall in blood pressure, particularly in giraffe in poor condition; and, once they collapse, death follows almost inevitably.

It is essential to administer a dose of narcotic that will not cause the animal to go down, as they experience great difficulty in rising in their bemused state and tire themselves so much in their efforts that they eventually succumb completely and die. In view of the risk of fatal hypotension, it is also essential to keep the animal on its feet. While walking about in a condition of 'twilight sleep', it can easily be roped and led into a crate mounted on the back of a low trailer. Once in the crate, the morphine-antagonist (100—200 mgm Nalorphine) is immediately injected into the jugular vein; and 100—200 mgm hydrocortisone (Vecortenol) and some 12,000,000 i.u. long-acting penicillin is administered intramuscularly. As soon as the animal's blood-pressure returns to normal, which may be deduced from the prominence of the Vena fascialis, the animal may be transported to the holding pen and released.

A completely safe and reliable drug-combination for young giraffe in the 600—1200 lb. class, is 2 mgm M–183, 50 mgm Hyoscine hydrobromide and 20 mgm Acetyl-promazine. The latter could probably be substituted with Fluanisone (20 mgm) with even better results. Instead of M–183 the potent analgesic Fentanyl (Janssen) at a dosage rate of 10 mgm/500 lb. body weight may also be used. For adult animals the dose of M–183 should be increased to 4 or 5 mgms.

If giraffe have to be transported over long distances, it is very advisable to keep them in a holding pen for some time until they are quite tame and will feed from the hand. Giraffe should be transported in crates large enough to allow the animal to lie down and rise without difficulty (preferably, a single animal in one large crate on the back of a 3-ton truck which is weighted down by a layer of sand ballast). Frequent stops should be made en route, and the animals allowed to rest to prevent fatal trauma in the extensor muscles of the front legs.

Over shorter distances giraffe may be transported in smaller crates, provided they are trussed in a special harness as is described by Riney and Kettlitz (1964).

Drugs of the Etrophine series, morphine and even diethylthiambutene hydrochloride (Themalon) are suitable for the capture of hippopotami (*Hippopotamus amphibius capensis* (Desmoulins)) on dry land, but are practically useless when the animals are in the water. The onset of the drug reaction is too rapid and the animals become completely immobilized, sink and drown.

The only drug combination which keeps the affected hippo buoyant for a sufficient length of time to allow a net to be brought in position and to haul it on to dry land before sinking is Sernylan and a suitable tranquilizer.

A Sernylan (Phencyclidine)-Chlorpromazine mixture gave very satisfactory results in the Kruger National Park and a number of hippo were successfully captured at a dosage rate of Sernylan 0·125—0·16 mgm/lb. and Chlorpromazine 0·25—0·4 mgm/lb. Chlorpromazine could profitably be substituted with Trifluopromazine, Acetyl-promazine or Azaperone.

## 2. Marking Techniques Employed to Date in the Kruger National Park.

Adjustable plastic collars, 6 in. broad, constructed of tough polyvinyl plastic strips stitched to a canvas backing and with a central celluloid support to prevent it from folding on itself, have been used to mark zebra, wildebeest, buffalo, waterbuck, kudu, impala, giraffe, tsessebe, and sable antelope for migration studies.

Different color combinations of white, black, yellow, and red have been used to indicate different marking stations and the colors remained distinguishable up to two years and longer. The collars have to be fitted tightly around the necks of zebra, and yet they still sometimes pull their heads through. Rival stallions also often tear the collars off when fighting. In horned antelope the collars can be fitted more loosely. Some have remained on for more than two years, during which time a great deal could be learned about migration routes, seasonal grazing grounds, etc. These plastic collars were clearly visible from the air. The manes and tails of zebra were cropped short to serve as further indicators of a marked beast.

Zebra and the larger antelope were also branded with special '8' numbered branding irons to provide a more permanent mark. Numbers branded on buffalo and wildebeest were clearly visible, but those on zebra were not so successful.

Plastic or metal ear tags and plastic streamers in the ears or tails of ungulates have also been used but are of limited practical value and are often lost by the animals.

Impala have been marked with plastic ear tags with attached streamers and with small plastic collars; and they have been branded with small wire branding irons.

The cropping of ears in a characteristic manner, as is the practice with small domestic stock, will probably prove to be the most practical method where large numbers of young impala are captured and marked for ageing studies.

Hot branding of elephants is not recommended, as the surface pigmented layer of the skin peels off and leaves an amorphous white scar.

Large (12 inches high) letters were drawn on the ears and rump of captured elephants with special hardboard stencils and grease pencil and painted over with several coats of a special, quick-drying poly-urethane marine plastic paint. These paints are highly durable and penetrate the cracks of the rough, bristly skin. Numbers which have been painted 6 months ago are still clearly visible, particularly if not obscured by mud. Numbers have also been engraved on the tusks with a hot soldering iron or electric drill to provide a permanent mark, and plastic streamers were fitted to the ears and tail. The latter are soon shed, however, and are of little use. A broad plastic band through slits in the tail skin will probably last quite a bit longer.

Baboons have been fitted with numbered, white ear tags and these lasted for a considerable time. Similar tags fitted to the ears of rhino and hippo were somehow shed by the marked animals within a few months.

The hair of dark-skinned animals can be bleached with peroxide and that of light-skinned animals blackened with silver nitrate solutions, but the results are not very satisfactory and the colored hairs are soon replaced.

## References

CAMPBELL H. & HARTHOORN A.M. (1963). The capture and anaesthesia of the African lion in his natural environment. *Vet. Rec.* **75**, 275–276.

FANN W.E. (1966). Use of methylphenidate to counteract acute dystomic effects of phenothiazines. *Am. J. Psychiat.* **122**, 1293.

HANKS J. (1967). The use of M-99 for the immobilization of the Defassa waterbuck (*Kobus defassa penricei*) *E. Afr. Wildl. J.* **5**, 96.

HARTHOORN A.M. (1967). Comparative pharmacological reactions of certain wild and domestic mammals to thebaine derivatives in the M-series of compounds. *Fed. Proc.* **26**, 1251.

HARTHOORN A.M. (1965). Application of pharmacological and physiological principles in restraint of wild animals. *Wildl. Monog.* **14**, p.78.

HARTHOORN A.M. & BLIGH J. (1965). New potent morphine analogues for the restraint of large hoofed animals. *Res. Vet. Sci.* **6**, 290–299.

HARTHOORN A.M. & PLAYER I.C. (1964). The narcosis of the white rhinoceros. A series of eighteen case histories. Proc. Fifth Intern. Symp. Dis. Zool. Anim. (1963). *Tyd Diergen.* **89**, 225–229.

HARTHOORN A.M. (1962c). Capture of white (square-lipped) rhinoceros, *Ceratotherium simum simum* (Burchell), with the use of the drug immobilizing technique. *Can. J. Comp. Med.* **26**, 203–208.

HIRST S.M. (1966). Immobilization of the Transvaal giraffe (*Giraffa camelopardalis giraffa*) using an oripavine derivative. *J.S. Afr. Vet. Med. Ass.* **37**, 85.

KING J.A. & CARTER B.H. (1965). The use of M-99 for the immobilization of the black rhinoceros (*Diceros bicornis*). *E. Afr. Wildl.* **3**, 19–27.

VAN NIEKERK J.W., PIENAAR U. DE V. & FAIRALL N. (1963b). A report on some immobilizing drugs used in the capture of wild animals in the Kruger National Park. *Koedoe* **6**, 126–134.

PIENAAR U. DE V. (1967). Operation 'Khomandlopfu' (Capture the Elephant). *Koedoe* **10**, 158.

PIENAAR U. DE V. (1967). The field immobilization and capture of hippotami (*Hippopotamus amphibius Linnaeus*) in their aquatic element. *Koedoe* **10**, 149.

PIENAAR U. DE V. (1968). Recent advances in the field immobilization and restraint of wild ungulates in South African National Parks. *J.S. Afr. Vet. Med. Assoc.* (in press).

PIENAAR U. DE V. (1968). The use of immobilizing drugs in conservation procedures for Roan antelope (*Hippotragus equinus equinus Desmarest*) *J. S. Afr. Vet. Med. Assoc.* (in press).

PIENAAR U. DE V. & VAN NIEKERK J.W. (1963). The capture and translocation of three species of wild ungulates in the Eastern Transvaal, with special reference to RO5-2307/B-5f (Roche) as a tranquilizer in game animals. *Koedoe* **6,** 83–91.

PIENAAR U. DE V. & VAN NIEKERK J.W. *et al.* (1966). The use of Oripavine hydrochloride (M-99) in the drug-immobilization and marking of wild African elephant (*Loxodonta africana* Blumenbach) in the Kruger National Park. *Koedoe* **10,** (in press).

PIENAAR, U. DE V. & VAN NIEKERK J.W. *et al.* (1966). Neuroleptic narcosis of large wild herbivores in the Kruger National Park with the new potent morphine analogues M-99 and M-183. *J. S. Afr. Vet. Med. Ass.* **37,** 227.

RINEY T. & KETTLITZ W.L. (1964). Management of large mammals in the Transvaal. *Mammalia* **28,** 189–249.

# 10

# Environmental Physiology

A T PHILLIPSON

It can be accepted that indigenous animals in their natural habitat are well adapted to their environment as otherwise they would not be there. The nature of their adaptation is a combination of behavioural and physiological characteristics, and a study of their physiology without a study of their behaviour is likely to be misleading if the object is to understand the place of a given species in any particular ecosystem. Even so, the ability of an animal to adapt itself to the changes in climate that occur within the 24 hours, from day to day and from season to season, is likely to be one of the many factors that influence its behaviour. Records of standard meteorological data of any area may be more informative about the conditions which an animal avoids rather than about the conditions with which it is in immediate contact. It is as important to study the 'microclimates' sought by animals, which may be altered by the mere fact of an animal being there, as it is to study the meteorological conditions generally prevailing.

The situation with domestic or captive animals is different, for these have to tolerate the conditions imposed upon them. Consequently, with domestic livestock a great deal of effort has been made to study their physiological responses to different environmental conditions. It is doubtful, however, whether the methods used may be of much assistance to the ecologist. Animal physiologists have constructed elaborate climatic chambers in which the temperature and the humidity and the air flow can be closely controlled and have made detailed observations of the effects of cold and hot climatic conditions on the animal. These studies define the degree to which the physiological mechanisms can protect the animal from climatic extremes, but the parameters established have to be translated to farm conditions with caution where animals are in groups, which, in turn, may influence the environment in which they are in immediate contact.

Climatic chambers are expensive; they demand a high degree of technical knowledge to construct and their use with tamed animals is of questionable value if the ultimate object is the study of animals in their free-living state.

If, on the other hand, the object is to select a given species of animal for domestication under fairly well-defined conditions, then there is reason to contemplate their use, although many simpler approaches are likely to be more helpful.

For the study of the environment of free-living animals, the first objective should be to obtain information on the meteorological conditions of the area and, from a knowledge of the behaviour of the species in question, to obtain information on the 'microclimates' which they choose for shelter and security. Information on their body dimensions, on the nature of their coats, and on their pigmentation of the hair and skin, on the density of the coat, on the presence, or not, of sweat glands and on the structure of the skin should be obtained. This information is basic to any further studies that may be contemplated. Restraint or tranquilization will be needed in order to measure relevant physiological parameters, and these in themselves may influence the results obtained.

Radio telemetric methods applied to animals are still insufficiently well developed to allow more than a comparatively short period of transmission over a severely restricted area on account of the weight needed to provide sufficient power for longer periods of transmission or a further radius of transmission. These methods can only be employed successfully in animals confined in a small area. With animals living at high altitudes, a study of the nature of their haemoglobin will be helpful.

The methods available for making such measurements are outlined, and their selection should be related strictly to the questions which are asked concerning free-living animals.

### The Measurement of Meteorological Variables

**Air temperatures.** Simple thermometers should be protected from direct and reflected radiation. It is usual to place them in a box with 'louvered' openings that will allow a good airflow. A rapid air flow is sufficient, by itself, to provide protection from radiation. Maximum and minimum thermometers should be used where possible. Clockwork thermographs depend upon the expansion and contraction of a coiled spring with temperature which moves a recording pen, writing on a slowly moving drum. Continuous records of temperature change can be recorded in this way.

**Humidity.** The direct measurement of water vapor pressure is seldom made. Instead, the relative humidity of the air at a given temperature is measured by recording the heat needed to evaporate water in the immediate vicinity of the bulb of a thermometer. The extent of evaporation depends upon the water vapor tension of the air, and cooling of the bulb is a measure of the latent heat needed for evaporation of the water. The relationships between the 'wet bulb' temperature, the 'dry bulb' temperature (the air temperature), the water vapor tension and the relative humidity of the air have been calculated for different barometric pressures; and a knowledge of two of the variables allows the remaining two to be read from Psychromatic charts.

The relative humidity is the value usually given in meteorological data and represents the water vapor tension of the air as a percentage of what it would be if the air was fully saturated at the given air temperature.

Two forms of 'wet bulb' thermometer are commonly used. The first is a permanent fixture. A muslin cover to the bulb is kept permanently moist by a wick passing to a water reservoir. Air is passed over the bulb at about 100 ft/min. by a fan operated by hand or by power. A convenient way of measuring the relative humidity is with a 'wet bulb' and 'dry bulb' thermometer fitted side by side in a device that is similar to the football fan's 'rattle.' The muslin surrounding the 'wet bulb' is moistened and the 'hygrometer' is whirled through the air. Readings from both thermometers then give the two measurements needed to read off the relative humidity from charts.

Clockwork thermographs adapted for 'wet bulb' recording can be used. The sensitive coil is kept moist permanently by a muslin cover that dips into water. A motor is needed to drive air across the muslin continuously. The coiled spring actuates a pen recording on a slow drum.

Humidity can also be measured by the contraction or relaxation of a human hair which again activates a writing pen. This instrument needs regular checking as the responses of a hair to water vapor tension change with time, especially when large variations in humidity occur. It is often combined with a thermograph recording changes in air temperature.

**Air movement.** This is usually recorded by a cup anemometer mounted on a vertical shaft. The rotations are recorded by means of a counter, but it can be made to work a pen recorder on a moving drum. The counter records revolutions over a period of time; pen recording gives also the variation throughout the 24 hours. The wind direction is shown by a vane attached to the top of the instrument.

The cup anemometer is not sensitive to air speeds of less than 3 m.p.h. (1·5 m/sec.). With fluctuating air speeds it overshoots when the speed drops and consequently overestimates the average air speed. It records only air movements in a horizontal direction. It should be placed away from buildings or trees that deflect air currents and may cause eddies to occur. It is usually mounted from 1 to 3 meters from the ground.

Air speeds at these distances from the ground are usually greater than they are a foot or so less from the ground, depending on the contours and ground cover. Direct measurements at different heights can be made, or alternatively the speed at different heights from the ground can be approximately calculated as follows:

$$V_1 = \frac{(0 \cdot 317 \log_r + 0 \cdot 366)}{(0 \cdot 317 \log_z + 0 \cdot 366)} V_2$$

when: $V_1$ = wind speed at $r$ feet from the ground
$V_2$ = wind speed at $z$ feet from the ground

For stable conditions the ratio should be squared.
For turbulent conditions the square root of this square should be used.

Measurements of air speeds at 4 feet, 2 feet and 9 inches from the ground have been made on the Scottish foothills by Cresswell and Thomson (1964). Four different sites were investigated and considerable variation between them was found in the relation between the air speed at 4 feet and at 9 inches. At the first site the air speeds at 9 inches from the ground varied from 72 to 130% of those at 4 feet from the ground. At the second site these values were 65 to 80%; at the third they were 37 to 55% and at the fourth, 27 to 52%. In only two measurements on the first site did the air speed at 9 inches from the ground exceed the air speed at 4 feet from the ground. The measurements made at 2 feet from the ground were much closer to those made at 4 feet on the first two sites but were approximately midway between the 4 feet and 9 inch measurements at the second two sites.

These data show that the calculation of air speed at different distances from the ground when it is measured at a standard distance may give very misleading results. In order to appreciate the wind speeds to which animals are exposed, measurements should be made at the height the animal is standing or lying. The position of the animal will influence the extent to which it is exposed.

The hot wire principle can be used to measure air flow as wind blown over a heated wire cools it and alters its electrical resistance, which can be measured. The response is instantaneous and it takes account of air movement in any direction. A source of power, however, is needed. This method may be too sensitive for general field use but may be adaptable to measuring air movement in 'microclimates.'

A simple technique of making individual measurements is to ignite a small quantity of gun powder from a standard container so that it produces a single puff of smoke. The time for this puff to travel over a measured distance is a measure of air flow. This method is suitable where low air velocities are encountered. A measure of air turbulence can be obtained by this method by determining the time taken for the puff of smoke to become invisible.

**Radiation.** Radiation may be direct solar radiation or radiation deflected by particles in the air or reflected from surrounding objects, or from the ground. It is not easy to measure the radiation that impinges upon animals, and their coat color and hair or wool density influences the amount of radiation they will absorb.

Direct solar radiation can be measured by a disc containing two concentric rings, one coated with white magnesium oxide and the other with lamp-black. The two rings contain the opposing junctions of a thermocouple. The disc is contained in a flat glass container set in a horizontal plane and protected from radiation except that received from the sky. The black ring absorbs radiation and the white one reflects it so that a temperature difference is set up according to the intensity of radiation, and this can be measured by connecting the leads from the thermocouple to a potentiometer. It records solar radiation and radiation reflected from clouds or particles in the sky.

Solar radiation (direct and reflected) can also be measured by a radiation meter similar to a photographic light meter but sensitive also to infra-red radiation. These are calibrated in g cal/sq. cm.

It will be appreciated that the intensity of direct solar radiation is influenced by the thickness of the atmosphere through which it passes and to the extent of cloud cover, dust, smoke, etc., that is in the atmosphere. The greatest intensity will occur at noon when the thickness of the atmosphere through which it passes is at its minimum. Observations on the time of day, latitude and cloud cover should be recorded with any measurements made.

It is not easy to measure reflected radiation as such from the ground or surrounding objects. One method that is used to assess reflected radiation is by a globe thermometer. This is a copper sphere 12·5 to 15·0 cm. in diameter which is fitted with a thermometer in its exact centre and painted black. This globe is shielded from the sun by a disc 17·5 cm. in diameter held 30 cm. away from the centre of the sphere. The shading disc needs adjustment at 15 min. intervals to allow for the change in position of the earth relative to the sun. It is necessary to know the air velocity and temperature of the surroundings to obtain a measure of the reflected radiation intensity. Prepared charts are available to relate globe temperature to air temperature and air velocity in ft/min. to give a corrected value to that actually recorded by the globe thermometer. The values are given as B.T.U/sq ft/hr. and one such unit is equivalent to 2·71 Kcal/sq m/hr.

**Rainfall and other forms of precipitation.** The simple rain gauge is a copper cylinder which has an aperture of precisely 5 inches in diameter. It is set firmly in the ground so that the aperture is exactly 12 inches above the surface of the ground and is level. The cylinder leads to a funnel which is set several inches below the aperture to prevent out-splashing. The water falling into the aperture runs into a graduated container set in the bottom of the cylinder. If rainfall is so heavy that the water overflows the measuring container, it spills into the bottom of the copper cylinder and can be withdrawn and measured.

The trapped water is measured at standard intervals of time. With snowfalls, the trapped snow is melted and measured as water.

Rain gauges that record are of the syphon type; with these the water from the aperture is led to a closed container in which is a float and a side arm syphon is attached. The container is balanced on a knife edge and when the float rises to its full extent it releases a catch that allows the container to fall to an angle. This sends a stream of water through the syphon so that the whole of the water syphons from the container as it regains its upright position. As the float rises, it writes by a pen recorder on a slowly moving drum; the number of times the container fills and empties in a given period is thus recorded on graduated paper and any residue in the container is read directly as a volume. In cold weather the apparatus has to be kept warm to prevent freezing.

**Measurement of 'microclimates'.** Measurements of temperature, humidity and air flow are the most relevant ones to attempt. The cover afforded by trees and undergrowth presumably can be measured by adopting the principles used for measuring these variables of the macroclimate, although the instruments will probably need modification if they are to be used in restricted spaces. It is a different proposition to measure the effects of the presence of an animal as any disturbance or unusual object may well deter the animal approaching its accustomed shelter. It should be possible to unobtrusively plant sensitive elements that record temperature and air flow and to connect them to recording instruments some distance away. The measurement of humidity with the usual instruments, however, will present difficulty owing to the necessity of maintaining a flow of air over the wet covering of a thermometer. Presumably the relaxation or contraction of a hair could be recorded electronically, but no existing instrument has been designed to do this. It should be possible to construct a sensitive thermocouple enclosed in a permanently moistened muslin cover, over which can be blown a gentle stream of air from a distance and without noise so that a 'wet bulb' temperature can be compared to a 'dry bulb' temperature from a similar air moistened thermocouple. The principal difficulty is for recording instruments to be set within reasonable reach of the elements in a manner that does not frighten away the animal under observation.

### The characteristics of the animal

The characteristics that are most relevant to the adaptability of animals to different environment can be divided into two categories; namely, those that are not distorted by the influence of temporary restraint and those that may be.

In the first category are the characteristics of the coat: the color, structure and distribution of the hair or wool; the pigmentation of the skin; the thickness of the skin and its structure; the distribution of sweat glands, if there are any; and the conformation of the body which should include measurements of weight, height, width, girth, leg length, tail length and measurement of the area of the skin, with special reference to those areas such as the ears and legs where heat exchanges can be rapidly controlled by vascular reactions.

In the second category are measurements of rectal temperature, skin temperature, heart rate, respiratory rate, the respiratory pattern, the transpiration of water from the skin, the gaseous exchanges of the animal and so on.

Little need be said of the measurements included in the first category that are essentially histological. It is probably best to collect and preserve specimens for later examination in a properly equipped laboratory. Otherwise, the two measurements that are likely to cause most trouble are the weight of the animal and its skin area. Portable machines that can weigh animals up to 500 lb. are on the market. Animals heavier than this present a problem. A new principle of weighing is needed for large animals which might possibly be based on the use of a portable platform which recorded electronically through strain gauges.

There is no universally satisfactory method for measuring the area offered by the skin. It has now been abandoned as a reference point for metabolic studies, and its place has been taken by the 'metabolic' weight of the animal which is expressed as the weight of the animal to the power of 0·75. The rationale of using the metabolic weight as a reference point is discussed by Brody (1946) and Kleiber (1965).

Measurements of the area of the skin from the point of view of tolerance to the environment may be useful if the object is to compare the area covering the trunk of the animal with the area covering the legs, tail, face, ears, or other appendages where heat exchanges are controlled to a greater extent and which do not have coat protection. Comparisons can only be approximate, as skin removed from the body is stretched and the degree of stretch affects the area subsequently measured. An instrument, based on the principle of a mileometer, has been used by Brody. It measures the skin area in strips in living animals and marks at the same time the area measured. Its use on the legs, head, ears, and tail, however, present considerable difficulties. The capacity of skins with different coverings and pigmentation to absorb or reflect radiation have been carried out on pelts exposed to solar radiation or to artificial radiation by spreading these over heat sensitive elements and recording the temperature transmitted through the skin. Use of hides or pelts in the reverse direction to determine their heat conserving capacity presumably could be undertaken although the measurements refer mainly to the properties of the structure rather than to their pigmentation. This kind of measurement may have comparative value but should be taken as being applicable only to the hide or skin as it is (Scholander *et al.*, 1950). In general, the insulating properties of the fur of animals is directly related to its thickness (Burton and Edholm, 1955).

**The measurement of physiological characters.** The measurement of the characters of the second category can only be done on animals that are tamed, restrained or tranquilized if the conventional methods are used. The introduction of telemetric methods of recording, however, allows some of these parameters to be measured in animals that are free within a paddock or a fenced-off area once the sensitive elements and the transmitting equipment are in place. Again this means that only animals that can be handled or narcotized with short acting agents can be used. It is questionable whether this kind of measurement comes within the sphere of field work, and there is no point in the field ecologist undertaking them until he is satisfied from his measurements of the environment and the habits of the animal that they will add materially to his understanding of the area and the animals within the area with which he is concerned.

Measurements of the insulation provided by the body, the coat or fleece may be also undertaken. The methods used with man and the principles involved are fully discussed by Burton and Edholm (1955) with respect to the measurement of tissue insulation and the insulation provided by clothing.

If the total heat produced by the body during a unit of time under standard conditions, the skin area, and the average skin temperature is estimated, then the temperature gradient between the deep body temperature and the skin divided by the heat loss per square meter of skin brought to a whole number can be taken as a measure of the insulation of the tissues. The same principles can be applied between the skin and the surface of the hair or wool covering and between the surface of the covering and the external air.

There are three difficulties in making these measurements. The first is the difficulty of measuring the surface of the skin which has already been discussed. The second is due to the fact that heat loss from the body consists of evaporative and non-evaporative heat losses, if the small heat losses due to excretion and expired air as such are ignored. If there were no evaporative heat losses from the body, then the heat loss of the animal should equal heat loss from the skin, from the covering of the skin to the air.

Evaporative heat losses from the body occur through the skin and the upper respiratory tract; and, unless evaporation through the skin is allowed for, heat flow from the body to the skin will be less than it should be and the apparent tissues insulation less than it should be. In man, under conditions when no sweating occurs, the unit for tissue insulation is multiplied by 1·21 to allow for evaporation loss through the skin.

The third difficulty is that skin temperatures from different areas of the animal may vary, especially when there is little or no coat cover. To get over this difficulty, skin temperature in representative areas have to be taken and the results integrated according to the extent of the areas concerned to give an average skin temperature. These methods have been used to study the internal and external insulation of the sheep under varying conditions of temperature and air speed by Joyce and Blaxter (1964) and for a comparison of the total insulation of different breeds under different climatic conditions. In these studies the environmental temperatures were maintained in the thermoneutral zone or well below the critical temperature for sheep in order to reduce interference due to evaporative heat losses to a minimum (Blaxter, Clapperton and Wainman, 1966).

**Adaptation to high altitudes.** The most relevant studies to make with animals that are acclimatized to high altitudes are on the oxygen carrying capacities of the blood. This will depend upon the number of erythocytes in circulation, their haemoglobin content, and the nature of the haemeoglobin. Relevant measurements are erythocytes counts, the packed cell volume, and the haemeoglobin value of the blood. Blood obtained from animals that are quietly resting and unalarmed is essential if data of value are to be obtained.

Determinations of the oxygen dissociation curve of blood is an essential guide to the suitability of haemoglobin types for high or low altitudes. In sheep, Harris and Warren (1955) found two electrophoretically distinct haemoglobins, and their distribution among sheep with different breeds suggested a geographic distribution (Evans, Harris and Warren, 1958). The haemoglobin that occurs with greater frequency in the northern type of breeds than in breeds of the middle east from its oxygen dissociation curve allows greater uptake of oxygen at low oxygen tensions than the haemoglobin associated with other breeds (Evans and Dawson, 1962). The carriage of oxygen by the blood and its availability to the tissues should be investigated in detail.

### References

BLAXTER K.L., CLAPPERTON J.L. & WAINMAN F.W. (1966). The extent of differences between six British breeds of sheep in their metabolism feed intake and utilization, and resistance to climatic stress. *Br. J. Nutr.* **20**, 283.

BRODY S. (1945). *Bioenergetics and Growth*. Reinhold Publishing Corporation, New York.

BURTON A.C. & EDHOLM O.G. (1955). *Man in a Cold Environment*. Monog. Physiol. Soc. 2. Edward Arnold (Publishers) Ltd., London.

CRESSWELL E. & THOMSON W. (1964). An introductory study of air flow and temperature at grazing sheep heights. *Emp. J. Exp. Ag.* **32**, 131.

EVANS J.V. & DAWSON T.J. (1962). Haemoglobin and erythrocyte potassium types in sheep and their influence on oxygen dissociation and haemoglobin denaturation. *Aust. J. Biol. Sci.* **15**, 371.

EVANS J.V., HARRIS H. & WARREN F.L. (1958). The distribution of haemoglobin and blood potassium types in British breeds of sheep. *Proc. Roy. Soc.* (*B*) **149**, 249.

HARRIS H. & WARREN F.L. (1955). The occurrence of electrophoretically distinct haemoglobins in ruminants. *Biochem. J.* **60**, xxix.

JOYCE J.P. & BLAXTER K.L. (1964). The effect of air movement, air temperature and infrared radiation on the energy requirements of sheep. *Br. J. Nutr.* **18**, 5.

KLEIBER M. (1965). Metabolic body size. Section 7 of *Energy Metabolism* Publ. No. 11, European Assoc. Anim. Prod. ed. K.L. Blaxter, Academic Press, New York.

SCHOLANDER P.F., WALTERS V., HOCK R. & LAWRENCE I. (1950). Body insulation of some arctic and tropical mammals and birds. *Biol. Bull.* **99**, 225.

*AIR MINISTRY METEOROLOGICAL OFFICE (1956). *Observer's Handbook*. H.M.S.O. London, M.O.

*DILL, D.B. ed. (1964). *Handbook of Physiology*, Sect. 4-Adaptation to the Environment. American Physiological Society, Washington.

*LEE D.H.K. (1953). *Manual of Field Studies on the Heat Tolerance of Domestic Animals*. F.A.O. Rome.


*These are not quoted in the text but are valuable sources of information on methods and instruments that are commonly used.

# 11

# Reproduction of Large Herbivores

## PART I. TECHNIQUES FOR INVESTIGATING REPRODUCTION OF FLOCKS OR HERDS UNDER FIELD CONDITIONS

### G R MOULE

### Reproduction in the female

The most useful information on overall reproductive performance will be obtained by obtaining age-specific data for the occurrence of oestrus, the ovulation and the fertilization rates, returns to service, and survival rates of young (Watson, 1957; Moule, 1960; Lamond et al., 1963). The discussion will be concerned with sheep; however, the methods and principles may be applied to other large herbivores.

**1. Breeding season.** The onset of the normal, basic rhythm of the oestrus cycle may be altered by the sight, sound, or smell of males (Watson and Radford, 1960; Lamond, 1962) or by stresses imposed on the females (Braden and Moule, 1962).

Sexual activity is not, by itself, a satisfactory indication of changes in the reproductive performance of ewes (Barrett et al., 1962).

Information can be obtained about the occurrence of oestrus and ovulation rates from the examination of the ovaries of live animals or of material obtained on post-mortem examination (Dun et al., 1960). Care is necessary is interpreting the results obtained from slaughter-house material because of the likelihood that stressful conditions imposed by transport prior to slaughter will induce ovulation (Braden and Moule, 1962).

Intepretations that have been placed upon the findings are:

(a) No large follicles and no corpora lutea: anoestrus

(b) Medium-size follicle: due to come on heat in a few days, or anoestrus

(c) Large follicle present: on heat or about to come on heat

(d) Small haemorrhagic spot on ovary: recently ovulated

(e) Mature corpora lutea: ewe recently on heat or pregnant
(f) Old corpus luteum: regressing corpus from an earlier cycle
(g) Cystic corpus luteum: recent stressful conditions.
Descriptions and methods for estimating the probable ages of corpora lutea are given by Dun *et al.* (1960).

**2. Detecting oestrus.** Oestrus may be determined with certainty only by the sexual behaviour of the female in the presence of the male. However, vaginal changes in cytology and volume accompany oestrus.

(*a*) *Sexual behaviour:* Females in oestrus may be identified by color marks left by entire or vasectomized males, on whose briskets crayon (Radford *et al.*, 1960) or mixtures of grease and pigment have been placed.
Inaccuracies may occur through:
  (i) Indistinct marks on the ewes due to inadequate mixtures of grease and raddle or through the choice of inappropriate crayons
  (ii) Mounting of females that are not on heat
  (iii) Lack of male libido

(*b*) *Vaginal cytology:* Oestrus is accompanied by changes in vaginal cytology and contents. During dioestrus the vagina is dry and the mucus membranes are blanched. The membranes become red and moist during oestrus and there is a copious flow of mucus (Radford and Watson, 1955).
Increasing amounts of oestrogens form the basis of the Allen-Doisey test (Allen and Doisey, 1923). Cells are collected from the vaginal wall by inserting a glass speculum to a depth of about 3 inches into the vagina of sheep. A small glass or stainless rod is then inserted through the speculum to a depth of 4—5 inches. After the rod has been rubbed on the side of the vaginal wall, the speculum is inserted to its full length, thereby lifting the wall away from the rod. The speculum and rod are withdrawn together and a smear is made by rolling the rod along the surface of a slide. A sterile speculum is required for each ewe.
*Staining.* The smear may be stained by Shorr's modification of Papanicolau method (Shorr, 1941) which helps differentiate cells by their color as well as by their shape. This method requires more time for the preparation of the slide, but time is saved in interpretation. For simplicity, Giemsa R66 (Gurr)* is more commonly used. The air-dried smear is covered with the

* G. T. Gurr, 136–144 New Kings Road, London SW6.

solution of Giemsa stain for from 4 to 7 minutes. The smear is then washed under running tap water until the blue colour disappears from the washings. When dry, the slide can be examined under a dry high-powered objective. A broad classification of the cells obtained from the vaginal wall and stained with Giemsa R 66 is:

| Type of cell | Description | Significance |
|---|---|---|
| *Epithelial cells* | | |
| Normal | Medium size, rounded deep continuous outline. Cytoplasm stains purplish-pink; nucleus stains deepest | *Negative* for oestrus |
| Squamous | Larger, thinner, flatter than normal cells, continuous outline, but angular. Cytoplasm stains fainter and bluer, granular nucleus larger, fainter, pinker than normal cells | Less than 40% squamous cells negative for oestrus |
| Cornified | Largest, thinnest, flattest of all cells. Angular, fragmented outline; cytoplasm stains pale blue; nucleus absent | *Positive:* Signs of cornification and at least 40% squamous cells* |
| | | *or* |
| Intermediate | Cells, intermediate stages between all three occur. | majority of cells squamous with some fully cornified |
| *Leucocytes* | | *or* |
| (white blood cells) | Small; cytoplasm stains light to dark pink. Distinct regular outline, variable shape, nuclei irregular | majority of cells fully cornified, remainder squamous. |

*'Signs of cornification and at least 40% of squamous cells' are criteria usually regarded as warranting further investigation before being classified as a firm 'positive'.

Slides which are interpreted as being positive may be divided into three grades:

I. at least 40% are squamous cells, some of which are cornified; normal cells and leucocytes present

II. majority of cells squamous, some fully cornified; a few normal cells and few leucocytes

III. majority of cells fully cornified, remainder squamous; few or no normal cells.

More detailed descriptions of interpretation, and a system of classification suitable for critical assay work, can be obtained from Robinson (1955). The Allen-Doisey test may also be used to confirm the presence of oestrogens in pasture (Sanger *et al.*, 1958). Therefore care is necessary when interpreting results obtained from the examination of changes in vaginal cytology of ewes grazing pastures that are potentially oestrogenic (Moule *et al.*, 1963).

(*c*) *Changes in vaginal volume.* When measured at a given pressure vaginal volumes change according to phases of the reproductive cycle (Wodzicka, 1960). During anoestrus the volume is at its lowest; it rises to a higher base level during the breeding season and shows marked increases at oestrus.

**3. Laparotomy.** In the ewe, laparotomy is essential for the diagnosis of early pregnancy, for spaying, egg transfer, and the determination of either ovulation rate or the precise time of ovulation. The work can be performed most conveniently if the ewe is suspended in a cradle (Lamond and Urquhart, 1961) or by a block and tackle. Shoulders should rest on the floor and hind legs should be about 1 ft. apart. If controlled in this position, it is advantageous to have an assistant standing behind the ewe to hold her flanks and to stretch the belly wall laterally. There should be little vertical tension on the ventral abdominal wall.

'Pentothal,'* 'Kemithal,'* or 'Nembutal'* may be used as a general anaesthetic, though the risks may be slightly greater with Nembutal if it is administered in haste. A combination of a tranquilizing drug, e.g. 'Largactil,'* together with a quick-acting local anaesthetic, is also satisfactory.

After the usual antiseptic precautions a paramedian incision $1\frac{1}{2}$—2 inches long is made just anterior to the udder and entrance can be made with one incision. To locate the uterus, the index and middle fingers should be directed towards the pelvic cavity in maidens or towards the abdominal cavity for parous ewes. The junction of the curved horns, or a horn itself, is easily recognized by touch and, if necessary, each horn can be withdrawn gently. There is danger of bursting a large Graafian follicle, or causing haemorrhages that lead to adhesions if tension is applied, or if the tract is handled roughly.

After the reproductive organs have been replaced, the incisions in the muscle and poritoneum can be closed with a mattress suture, and those in the skin by Michel clips.

*Pentothal—Abbott Labs; Kemithal—I.C.C.; Nembutal—Abbott Labs; Largactil—May & Baker.

**4. Time and number of ovulations.** Unless the laparotomy incision is large, it is preferable to withdraw the ovaries separately and then note the number, size and state of the follicles or of the corpora lutes. Special attention must be given to the poles of the ovaries, which may easily be obscured by the attachments with the broad ligaments.

**5. The diagnosis of pregnancy.** Failure to return to oestrus after service has often been used to diagnose pregnancy. Errors may arise from:
   (a) service of females not in oestrus,
   (b) the return of oestrus by pregnant females (Williams *et al.*, 1956),
   (c) the occurrence of anoestrus after an infertile service.

The most accurate way of diagnosing pregnancy in ewes within 6 to 8 weeks of service is to perform a laparotomy and to palpate the cervix and uterus. The incision needs to be large enough to admit only the second finger. Pregnancy may be diagnosed from the degree of swelling of the uterine horns and tension on the cervix. After six weeks' gestation, cotyledons may be felt. In cattle, pregnancy may be diagnosed by *per rectum* examination.

**6. Determining embryonic loss.** If during the breeding season females do not come on heat after the lapse of one oestrus cycle following service but do so after two or three cycles, there is circumstantial evidence that embryos are being lost. This may be investigated in ewes by slaughtering 48 hours after service. A further group is slaughtered 28 days after service, and the remainder are allowed to complete pregnancy under surveillance. Comparisons can be made between:
   (a) the number of corpora lutea and fertilized ova recovered from the first group slaughtered
   (b) the number of corpora lutea and embryos recovered from the second group slaughtered
   (c) the number of foetuses carried to full term as compared with the estimates of eggs shed, eggs fertilized, and embryos implanted.

The embryonic loss can be calculated also from data on the number of females that:
   1. mate and do not return to oestrus within 40 days but fail to lamb (N)
   2. mate and produce young normally (L)

The percentage of embryonic loss is calculated from the formula

$$\frac{N}{(N + L)} \times 100$$

The accuracy of the method depends on the assumption that anoestrus does not supervene within 40 days of the commencement of joining, and on the accuracy with which oestrus is detected, the failure of oestrus to occur among pregnant ewes, and the number of twin ovulations.

**7. Measuring lactation.** Three methods have been used to measure lactation:
(a) the young are separated from their mothers for given periods and are then weighed before and after sucking
(b) the milking by hand or milking machine after the intravenous injection of 5 i.u. of posterior pituitary extract containing oxytocin
(c) milking morning and evening.

The first method measures the milk intake of lambs and not necessarily the rate of milk production; the last two methods measure the rate of secretion over a given period and may be used to find the total potential for milk production and the shape of the potential lactation curve. These methods should give reasonably similar results, except where the appetite of the lamb is less than the milk production of the ewe. The rate of lactation may change rapidly at its onset of lactation or when it is curtailed suddenly as a result of adverse environmental conditions (Moule and Young, 1961), parasitism or ill health, thereby causing divergent results.

The separation of ewes from their lambs may interfere with milk production by creating a state of anxiety in mothers. To overcome this, Owen (1953) placed covers over the teats of ewes to prevent sucking. The ewes and their lambs were allowed to graze together and the covers were removed from the udders every few hours. The lambs were weighed before and after sucking. Most assessments of the milk consumption of lambs have been based on records obtained over one 24-hour period each week, but for reasons outlined by Moule and Young (1961) these estimates may not be very accurate if the weather is hot. Other workers (Barnicoat *et al.*, 1949a, 1949b) have drawn attention to the high correlation between milk consumption of lambs and their gain in weight during the first weeks of life. The exact terms of this relation should be noted and not misconstrued as being a measure of the ewe's capacity to produce milk.

The rate at which milk is secreted can be measured by hand milking (McCance, 1959) or by milking machine. The ewe is restrained in a normal position by placing her legs and udder in holes cut in a canvas stretcher lifted high enough to remove the animal's feet from the ground. The intravenous injections of oxytocin are given to ensure the let-down of milk and milking is continued until the udder is completely empty. The lambs are precluded from sucking for the ensuing 1—4 hours so that the milk is allowed to accumulate. The udder is again emptied after oxytocin injections which may have to be repeated to reach a satisfactory end point. Satisfactory yields can be obtained by careful handling of animals not treated with oxytocin but milked by hand in the manner normally used for commercial production.

### Reproduction in the male

An assessment of the reproductive performance of the male can be obtained from the numbers of females served, the numbers that conceive, and the numbers of services per conception. The number of services in a given time and the provocative time have been used to measure the libido (Anderson, 1945). Libido is not a useful indication of either fertility or fecundity.

Palpation of the scrotum and its contents can give information only about clearly established clinical conditions.

**1. The collection of semen.** One of three methods may be used to collect semen:

(a) **Electro-ejaculation.** An apparatus consisting of a small vibrator and transformer, capable of delivering an alternating current of up to 120 milliamperes, 60 volts and 50 cycles, is used to cause ejaculation in rams (Gunn, 1936). A bipolar electrode is described by Blackshaw (1954) who used stimuli applied initially for 15, 10, and 5 seconds respectively with further short stimuli if necessary. Other workers find that stimuli of 8, 5 and 3 seconds duration are satisfactory. The glans penis is extruded from the prepuce, surrounded by cotton gauze and held so that the urethral process hangs freely in the mouth of a 5/8 inch diameter, graduated, glass centrifuge tube. After each stimulus has applied, the penis is massaged anteriorly to expedite the flow of seminal secretions. An end point is reached when the secretions are clear.

(b) **Service into an artificial vagina.** The use of the artificial vagina has been described by Anderson (1945) and by Frank (1950). The casing of the vagina can be made from standard polythene tubing with an external diameter of 63 mm. for rams. The ends of the rubber lining fold back over the ends of the casing, and care is necessary to avoid wrinkles and folds down the lumen. Water at 50—60° C. is added to the chamber between the lining and the outer casing to about quarter-capacity. Soft medical paraffin is smeared in the first 2—3 in. of the lumen. The wide-mouth collection tube is inserted in the other end of the lumen and is held by pressure when the water chamber is inflated by blowing air through the tap. Sufficient inflation has been achieved when slight pressure is exerted on a finger inserted into the lumen. The optimum temperature within the lumen is 41° C $\pm$ 1° C. Training is required and the detailed procedure has been documented by Salamon and Lindsay (1961).

Numerous collections can be made from a male in one day. The use of artificial vagina is preferable to an electro-ejaculation if the work includes studies on potential semen production, as 'exhaustion tests' can be undertaken. The results obtained from exhaustion tests will give some indication of the sperm reserves in the ampula and epididymes without slaughtering the experimental animal.

(c) **Service into vagina.** Semen can be recovered from the vagina immediately after service (Terrill, 1940). The vulva is cleaned and the vagina should be emptied of mucus. Immediately after service, a glass speculum is inserted into the vagina and the semen withdrawn with a pipette and then discharged into a centrifuge tube. The volume obtained may be affected by vaginal mucus and by the difficulty of collecting all the semen. Contamination of the semen by vaginal cells is inevitable.

**2. Evaluation of semen.** Many suggestions have been made for evaluating semen, but no single test gives a reliable indication of fertility.

(a) *Volume of semen sample.* This is read directly from the graduated collection tube. Semen adheres readily to glass and, as the meniscus is usually high, it should be read to its base.

(b) *Concentration of sperm.* Four methods are in common use:
    (i) *Visual assessment of color and consistency.* Semen samples can be classified as being thick-creamy, creamy, milky, cloudy (opaque), or clear. It

has been suggested that the approximate number of sperm which may occur in samples of ram semen of different consistency are as follows (Gunn *et al.*, 1942):

| Classification | Approximate no. sperm (millions per cc) |
|---|---|
| Thick-creamy | More than 3000 |
| Creamy | 2000—3000 |
| Milky | 500—2000 |
| Cloudy (opaque) | Less than 500 |
| Clear | Insignificant |

(ii) *Comparison with barium sulphate standards.* The preparation of barium sulphate standards has been described in detail by Comstock *et al.* (1943). A portion of each semen sample is diluted 1/20 to 1/250 using physiological (0·9%) saline and is mixed by inversion ten times in a tube of the same diameter as the standards. The diluted semen sample is then compared with the standard tubes of barium sulphate using the clarity of a wire grid held behind the tubes and not the color of the suspension for comparison. Cellular debris and leucocytes affect readings, but damaged or abnormal sperm do not affect accuracy.

(iii) *Haemocytometer counts.* In the improved Neubauer ruling an area of 1 sq. mm. is divided into 25 large squares by sets of triple lines. Each large square is subdivided into 16 small squares by a single line. The depth of the film of fluid trapped beneath the cover-slip is 0·1 mm., so the volume of fluid held above the 15 large squares is 0·1 c mm. (0·0001 cc).

A sample of the semen is diluted with formolsaline according to visual assessment as follows: creamy 1/500, milky 1/200, cloudy 1/100. These dilutions are made with a Thoma pipette or a volumetric flask and graduated pipette may be used.

The haemocytometer chamber is flooded, and the sample is allowed to settle for 4—5 minutes before it is counted under a magnification of × 240— 300. Sperm in sixteen small squares within each of five large squares are counted. If the total count of sperm is less than 100, additional squares should be counted. Sperm heads only are counted; those touching the top and right-hand lines are included, while those on the bottom and left-hand lines of the large squares are excluded.

(iv) *Turbidometric comparisons* are based upon the increasing impedance to the passage of light with increasing sperm numbers in a nephelometer, photocolorimeter, or spectrophotometer.

The readings for the instruments are standardized against haemocytometer counts. Optical density can be plotted against sperm numbers, the regression calculated, and the standard errors placed upon estimates.

**3. Total sperm numbers.** The total numbers of sperm per ejaculate are calculated by multiplying the sperm density per ml. by the volume of the ejaculate in ml.

**4. Motility.** Motility should be assessed soon after the semen is collected. If an interval of more than a few minutes elapses, the samples should be kept in a container warmed to 36° C.

A drop of semen is placed on a warmed slide on a microscope stage with the temperature controlled at 36° C. The slide should have a thin glass bar cut from coverslips cemented to it, thereby making three or four divisions. A drop of semen is placed on the slides, between the bars on which a coverslip is placed. The motion is usually graded as follows:

| Grade | Motion |
|-------|--------|
| 0 | No currents |
| 1 | Few slow currents or waves |
| 2 | Many moderate waves |
| 3 | Many sweeping waves |
| 4 | Numerous vigorous waves |
| 5 | Numerous rapid and vigorous waves, pattern tumultuous |

This assessment is subjective and practice is required to achieve repeatability.

**5. Proportion of normal sperm.** The morphology of stained sperm can be examined under dry high-power or oil immersion, and that of unstained sperm by phase-contrast microscopy (Austin and Bishop, 1958).

A smear is prepared for staining by placing a drop of semen near one end of a clean, warmed slide. A second warm, clean slide is placed face down on the first slide, and the semen is allowed to spread between the two. The slides are then drawn apart lengthways with a smooth, quick motion, with their faces parallel and with the least pressure between the slides. The smear is then air dried quickly or given any special treatments that different staining techniques may demand. Among the stains that have been used are haemalum and eosin (Cary and Hotchkiss, 1935), carbolfuchsin and analine gentian violet (Starke, 1949), carbofuchsin (McKenzie and Berliner, 1937) and opal blue (Langerlof, 1936).

The morphological abnormalities most commonly observed are taillessness, coiled tails, and defective staining. Kinks in the middle pieces, enlarged middle pieces, double middle pieces and tails, protoplasmic droplets in the middle pieces, filiform middle pieces, and pyriform narrow, or double heads are uncommon.

An examination should always be made for the presence of pathogens in semen, especially as palpation of testes is unlikely to detect the voiding of infective organisms (Edgar, 1959).

**6. Proportion of live sperm.** The proportion of live sperm may be determined by differential staining using nigrosine/eosin. The stain is prepared by dissolving 5 gram of Eosin Y (Gurr) in 3·00 ml. of 5% nigrosine solution made with distilled water (Campbell *et al.*, 1956).

Seven or eight drops of stain are placed in each of a number of small test tubes in a water bath 30° C. One drop of freshly obtained semen is placed carefully in each tube, and the semen and stain are mixed by shaking the tube back and forth. The mixture and stain are allowed to stand for 5 minutes in the water bath at 30° C. If more than 2 or 3 minutes are likely to elapse before the collection of semen and the commencement of staining, the temperature of the semen should be maintained at 30° C.

A pipette is flushed with a mixture of semen and stain. The mixture is drawn into the pipette; the first drops are discarded, and one is placed on one end of the slide. The slide is air dried and examined at a magnification of × 300 or more.

The live sperm do not take up stain; their outlines are clearly defined and their bright, refractile heads are in sharp contrast with the dark background of nigrosine. The dead sperm stain pinkish; their outlines are not so clearly defined and the head is not in sharp contrast with the background. The percentage of living sperm is estimated from counts involving 200 or more sperm. The work is facilitated by using a deep-blue filter on the light stage.

**7. Vasectomy.** Vasectomy renders males infertile without impairing their sexual activity. A few non-motile sperm may appear in the seminal plasma for years after vasectomy, but they are of no significance. The simple procedure for adult rams is as follows: apply a local anaesthetic, fix the position of the vas deferens with pressure between the forefinger and the thumb and hold it away from the cord and the veins. One firm incision is made down onto the vas which erupts through the incision and can be recognized by its firmness and glistening white appearance. Two inches of the vas can then be excised and the skin wound closed with a single Michel clip.

Alternatively, an incision may be made through the mid-neck of the scrotum after the area has been anaesthetized. The incision should go through the skin and the sub-cutaneous connective tissue to permit the insertion of a female urethral probe or similar blunt instrument. The probe is inserted downwards from the lateral aspect of the incision and passes between the skin and the spermatic cord. The latter is raised and the vas can be recognized through the connective tissues from which it is freed and excised. The skin wound can be closed with clips or interrupted sutures.

### References

ALLEN E. & DOISEY E.A. (1923). An ovarian hormone. *J. Am. Med. Ass.* **81,** 819–921.

ANDERSON J. (1945). The semen of animals and its use for artificial insemination. Tech. Com. Imp. Bur. Anim. Breed. and Genet., Edinburgh.

AUSTIN C.R. & BISHOP M.W.H. (1958). Some features of the acrosome and perforatorium in mammalian spermatozoa. *Proc. Roy. Soc. Bull.* **149,** 234–240.

BARNICOAT C.R., LOGAN A.G. & GRANT A.I. (1949a). Milk-secretion studies with New Zealand romney ewes. Parts I and II. *J. Ag. Sci.* **39,** 44–45.

BARNICOAT C.R., LOGAN A.G. & GRANT A.I. (1949b). Milk-secretion studies with New Zealand romney ewes. Parts III and IV. *J. Ag. Sci.* **39,** 237–248.

BARRETT J.F., REARDON T.F. & LAMBOURNE L.J. (1962). Seasonal variation in reproduction performance of Merino ewes in northern New South Wales. *Aust. J. Exp. Ag. Anim. Husb.* **2,** 69–74.

BLACKSHAW A.W. (1964). A bipolar rectal electrode for the electrical production of ejaculation in sheep. *Aust. Vet. J.* **30,** 249–250.

BRADEN A.W.H. & MOULE G.R. (1962). The induction of ovulation in anoestrous ewes. *Aust. J. Exp. Ag. Anim. Husb.* **2,** 75–77.

CAMPBELL R.C., DOTT H.M. & GLOVER T.D. (1956). Nigrosin eosin as a stain for differentiating live and dead spermatozoa. *J. Ag. Sci.* **48,** 1–8.

CARY W.H. & HOTCHKISS R.S. (1935). A differential stain that advances the study of cell morphology (semen). *Cornell Vet.* **25,** 79–80.

COMSTOCK R.E., GREEN W.W., WINTER L.M. & NORDSKOG A.W. (1943). Studies of semen and semen production. *Univ. Minn. Ag. Exp. Sta. Tech. Bull.* 162.

DUN, R.B., ALMED W. & MORRANT A.J. (1960). Annual reproductive rhythm in merino sheep related to the choice of a mating time at Trangie, central western New South Wales. (1960). *Aust. J. Ag. Res.* **11,** 805–826.

EDGAR D.G. (1959). *N.Z. Vet. J.* **7,** 61–63.

FRANK A.H. (1950). Artificial insemination in livestock breeding. *U.S. Dept. Ag. Wash. D.C. Cir.* 567.

GUNN R.M.C. (1936). Fertility in sheep. Artificial production of seminal ejaculation and the characters of the spermatozoa contained therein. *C.S.I.R.O. Aust.* **94**, 1–116.

GUNN R.M.C., SANDERS R.N. & GRANGER W. (1942). Studies in fertility in sheep. II. Seminal changes affecting fertility in rams *C.S.I.R.O. Aust.* **148**, 1–140.

LAMOND D.R. (1962). Anomalies in onset of oestrus after progesterone suppression of oestrus cycles in ewes associated with introduction of rams. *Nature* **193**, 85–86.

LAMOND D.R. & URQUHART E.J. (1961). Sheep laparotomy cradle. *Aust. Vet. J.* **37**, 430–431.

LAMOND D.R., WELLS K.S. & MILLER S.J. (1963). Study of a breeding problem in merino ewes in central Queensland. *Aust. Vet. J.* **39**, 295–298.

LANGERLOF N. (1936). Sterility in bulls. *Vet. Rec.* **48**, 1159–1173.

MCCANCE I. (1959). The determination of milk yield in the merino ewe. *Aust. J. Ag. Res.* **10**, 839–853.

MCKENZIE F.F. & BERLINER V. (1937). The reproductive capacity of rams. *Res. Bull. Mo. Ag. Exp. Sta.* 265.

MOULE G.R. (1960). The major causes of low lamb marking percentages in Australia. *Aust. Vet. J.* **36**, 154–159.

MOULE G.R., BRADEN A.W.H. & LAMOND D.R. (1963). The significance of oestrus in pasture plants in relation to animal production. *Anim. Breed. Abstr.* **31**, 139–157.

MOULE G.R. & YOUNG R.B. (1961). Field observations on the daily milk intake of merino lambs in semi-arid tropical Queensland. *Qnsld. J. Ag. and Anim. Sci.* **18**, 221–229.

OWEN J.B. (1953). Milk yield of hill ewes. *Nature* **172**, 636–637.

RADFORD H.M. & WATSON R.H. (1955). Changes in the vaginal content of the merino ewe throughout the year. *Aust. J. Ag. Res.* **6**, 431–445.

RADFORD H.M., WATSON R.H. & WOOD G.F. (1960). A crayon and associated harness for the detection of mating under field conditions. *Aust. Vet. J.* **36**, 57–66.

ROBINSON T.J. (1955). Qualitative studies on the hormonal induction of oestrus in spayed ewes. *J. Endoc.* **12**, 163–173.

SALAMON S. & LINDSAY R.D. (1961) In: *Proc. Conf. on Artificial Breeding in Sheep in Australia*, ed. E.M. Roberts. Univ. N.S.W., Sydney. 173–182.

SANGER V.L., ENGLE P.H. & BELL D.S. (1958). The vaginal cytology of the ewe during the estrous cycle. *Am. J. Vet. Res.* **19**, 283–287.

SHORR E. (1941). A new technic for staining vaginal smears: III, a single differential stain. *Sci. (N.S.)* **94**, 545–546.

STARKE N.C. (1949). The sperm picture of rams of different breeds as an indication of their fertility. II. Rate of sperm travel in the genital tract of the ewe. *Onderstepoort J. Vet. Sci. Anim. Ind.* **22**, 415–525.

TERRILL C.E. (1940). Comparison of ram semen collection obtained by three different methods for artificial insemination. *23rd Ann. Proc. Am. Soc. Anim. Prod.* p. 201.

WATSON R.H. (1957). Wastage in reproduction in merino sheep. *Aust. Vet. J.* **33**, 307–310.

WATSON R.H. & RADFORD H.M. (1960). The influence of rams on onset of oestrus in merino ewes in the spring. *Aust. J. Ag. Res.* **11**, 65–71.

WILLIAMS S.M., GARRIGUS U.S., NORTON H.W. & NALBANDOV A.V. (1956). The occurrence of estrus in pregnant ewes. *J. Anim. Sci.* **15**, 978–983.

WODZICA M. (1960). Changes in the dilatability of the ovine vagina in relation to oestrus and anoestrus. *Aust. J. Ag. Res.* **11**, 570–575.

# PART II. PHOTOPERIODICITY IN REPRODUCTION: COLLECTION OF FIELD EVIDENCE

### N T M YEATES

There is a wealth of experimental evidence indicating the importance of the daily photoperiod in regulating the periodicity of reproduction in seasonal breeding species.

Thus 'light reversal' experiments conducted in light chambers have shown beyond doubt that the seasonality of reproduction in such ungulates as the sheep, goat, and horse is photoperiodically regulated. Many other species (including avians) are similarly influenced, and experiments with certain of them have led to a general acceptance that the seasonal gonadal responses to day-light are brought about by a retino-pituitary relay which is partly neural and partly humoral. Hypothalamic neurosecretion has traditionally been regarded as constituting part of this pathway; however, species differences in the location of the secretory cells have not been defined and it is not clear to what extent other structures, such as the pineal body, might also be involved.

While the precise details of mechanism will only be solved by sophisticated laboratory studies, the question which arises in the present context is how field evidence of photo-periodicity may be obtained among animals which have not been domesticated and which have evaded the attention of investigators.

A number of useful leads may be obtained from *field observations*.

**1. Species which have evolved in high latitudes.** Adapted as they are to large seasonal light differences, the high latitude species naturally provide the most numerous instances of photoperiodic influence. However, by no means all the species of such latitudes breed seasonally and classification of them on this basis is a first requirement.

Statistics on the reproductive state (e.g. pregnancy, ovulation, seminal and male tract activity) of trapped or killed individuals on a monthly basis throughout the year will facilitate this. Then, in the case of proven seasonality, the periodicity of breeding should be plotted using the calendar months as a time base and compared with similar records on successive years. In general, precise repetition of the annual breeding pattern on a monthly basis strongly indicates photoperiodic regulation, while lack of repetition suggests the influence of other exteroceptive factors less regular than day-light change. In fact, however, instances of regulation of seasonal breeding by factors other than light are rare among the high latitude species.

**2. Species evolved near the equator.** The tropical species present a more complicated picture. For although a great many (probably the majority) of them are strictly nonseasonal in their breeding habits, variable periods of anoestrus do occur, and this may be difficult to differentiate in the field from a truly periodic anoestrum which, by definition, should be predictable in terms of the calendar months, each year.

It might be thought that the relative uniformity of the tropical light environment would rule out the likelihood or even the possibility of photoperiodic stimulation. While this may be so for equatorial latitudes, it is not so for the fringes of the tropics where seasonal differences in daylength are appreciable. Even in the New Hebrides (15° S. lat.), where the summer-winter light difference is only 1¾ hr. (and where the environment is otherwise so uniform), the bat *Miniopterus australis* commences breeding at the same time, almost to the day, each year. It is hard to attribute such clearly defined sexual periodicity to anything but photo-stimulation; however, final proof must await light-chamber studies. (Perhaps it will then also be found that *Miniopterus* is specially well-equipped visually for light perception—a suggestion which is supported by the reported precision with which it emerges from its day-time caves in New Hebrides, at exactly the same time before sunset each evening.)

In summary then:

(i) Some attempt should be made to determine the approximate latitude of origin of the species under investigation.

(ii) If the periodicity of sexual function is clearly defined and highly repetitive annually in terms of the calendar months, photostimulation should be suspected.

(iii) Alternation of breeding activity and quiescence which follows no exact pattern of annual repetition suggests either an interaction of light with other stimuli (e.g. temperature, rainfall, nutritional state), or operation of a stimulus other than light.

(iv) Absolute equality of breeding intensity in all months of the year eliminates light and other environmental factors from consideration; (this is the usual situation among tropical species, assuming normal health and nutrition).

## References

YEATES N.T.M. (1947). Influence of variation in day length upon breeding season in sheep. *Nature* **160**, 429.

YEATES N.T.M. (1954). Daylight changes, Chapter 8, in *Progress in the Physiology of Farm Animals*, ed. Hammond J., Butterworths, London.

YEATES, N.T.M. (1965). Modern aspects of Animal Production. Butterworths, London.

## PART III. COLLECTION AND FIXATION OF FIELD SPECIMENS FOR HISTOLOGICAL EXAMINATION

### P N O'DONOGHUE

Ideally, as soon as collected, material should be divided into the small pieces that histologists love but so seldom receive. This is to permit fixatives, despite their relative reluctance to penetrate a block of tissue, to reach all parts before much autolysis can occur. Optimum size of such pieces will vary with the specimen. Generally, at least one dimension should not exceed a centimeter; and, if possible, the block should present plenty of cut surfaces because fluids will pass more easily across a cut face than across an intact epithelium.

However, unlike material derived from experimental animals, specimens collected in the field are frequently of unknown potential. The exact parts of an organ, for example, that will prove of interest and the techniques that will have to be applied may be totally unpredictable. Thus, such specimens will tend to be bulky; and the main object must be to preserve them as simply as possible to permit the subsequent application of more elegant techniques in more convenient circumstances.

The fixative of choice in such a situation is formaldehyde, carried as a concentrated (40% w/v) solution and diluted with water to 5—10% formaldehyde (12—25% of the 40% solution) when required. This is straightforward and, although unpleasant, is less subject to evaporation than alcohol and less embarrassing if spilled than acids or flammable fluids. Further, it has the great virtue of doing comparatively little beyond preserving, and thus leaves available the widest possible range of future techniques—secondary fixation, fat study, various embedding methods—more so than any other practical preservative except deep freezing.

The problem of fixative penetration, or lack of it, which is solved in the laboratory by division of large specimens, may be largely overcome in the field by *slashing* (partial division). Where this is undesirable, *perfusion* through the blood vessels should be considered; and, in general, field workers would be far better advised to carry a few syringes, probing needles, cannulae and haemostat forceps than to burden themselves with a bulky selection of fixatives and supernumerary specimen jars.

The most obvious and desirable refinement to this suggested procedure is to carry sodium chloride and prepare formol-saline instead of straight aqueous formaldehyde, and to permit a solution of $0.2\%$ w/v (very roughly) sodium nitrite in saline ($0.85\%$ w/v) to be used to flush and dilate blood vessels prior to perfusing with fixative. There are many other improvements that could be made; but they will not only burden an expedition, they will also limit severely the delegation of collecting tasks to scientifically unpractised helpers (trappers, gamekeepers, etc.). The extent to which a field expedition can be made into a travelling laboratory must vary from case to case.

A wide selection of polythene bags and appropriate plastic clips is available, and specimens can be individually bagged in fixative with their labels. Bags of 120 gauge sheeting about 3 in. $\times$ 4 in. (8 $\times$ 10 cm.), 7 in. $\times$ 10 in. (18 $\times$ 25 cm.) and 10 in. $\times$ 17 in. (25 $\times$ 43 cm.) will be suitable for most sizes of specimens. Spinous or heavily clawed animals are better in jars. Bags may be labelled with a felt-pen, the pigment of which dissolves in organic solvents but not readily in 10% formaldehyde. The bags may be closed with plastic rachet clips (Fisons Scientific Ltd., Loughborough, Leicestershire, England). A number of those bags may then be contained in one large glass or plastic vessel which is also charged with fixative. The double layer of fluid cushions specimens during transit and guards against desiccation where field temperatures are high and organization rather primitive. Particular care should be taken to select labels that do not disintegrate in the fixative and writing materials that do not fade or dissolve. Pencil on stout paper (e.g. luggage labels) will usually suffice. It is well worth writing on both sides of the label.

## PART IV. VITAL STATISTICS AND MEASUREMENTS OF REPRODUCTION RATE IN SHEEP

H Newton Turner

The zoography of the sheep may be used to illustrate the vital statistics of mammalian populations in conditions where an adequate census of age, sex and fecundity is possible. Findings from different sources sometimes cannot be compared because of differences in presentation—the number of lambs born per ewe lambing, for example, cannot be related to number of lambs born per ewe joined (put to the ram) unless the number of ewes which failed to lamb is also known. If some degree of standardization can be introduced into reporting, comparisons will be simplified.

A basic equation is presented which can be expanded to cover most aspects of vital statistics likely to be required, and which lends itself to manipulation for extracting various types of information:

$$L_{X_i J} = E_{PJ} \times L_{BP} \times L_{X_i B} \quad \ldots \ldots \ldots \quad (1)$$

where $L_{X_i J}$ = number of lambs surviving to age $X_i$, per ewe joined

$E_{PJ}$ = number of ewes lambing, per ewe joined

$L_{BP}$ = number of lambs born, per ewe lambing

$L_{X_i B}$ = number of lambs surviving to age $X_i$, per lamb born.

The terms $E_{PJ}$ and $L_{X_i B}$ can be expanded further to cover losses at various stages, the full version being:

$$L_{X_i J} = \pi(1 - E_{Y_p Y_q}) \times L_{BP} \times L_{X_k X_l} \quad \ldots \ldots \quad (2)$$

where $\pi$ stands for continued product, $E_{Y_p Y_q}$ the proportion of ewes which fail at each stage of the period from joining to lambing, and $L_{X_k X_l}$ the survival rate of lambs in different stages of the period from birth to age $X_i$.

The term 'ewes joined' should refer to all ewes put to the ram, in the case of natural matings, or to all ewes presented for the identification of oestrus, in the case of hand service or artificial insemination. Frequently the conception rate, based on ewes inseminated, is quoted for artificial insemination; this figure measures the success of the insemination procedure, but gives a biased estimate of the productivity of the whole flock, unless the number of ewes failing to come on heat can also be included.

The term $L_{BP}$ should include all lambs born, whether dead or alive, the term $L_{X_k X_l}$ being used to measure survival; thus if $k = 1 = 0$, $L_{X_o X_o}$ represents lambs born alive, as a proportion of all lambs born.

As an example, suppose we have 100 ewes joined, of which 2 fail to mate, 3 fail to conceive, and 4 lose their embryos; the 91 ewes which lamb bear 105 lambs; of these 105 lambs, 6 are dead at birth, 10 die before the end of the lambing period and another 4 before weaning. We then have, if W stands for weaning:

$$L_{\underset{w}{X\ J}} = (1 - \frac{2}{100})\,(1 - \frac{3}{98})(1 - \frac{4}{95}) \times \frac{105}{91} \times (1 - \frac{6}{105})\,(1 - \frac{10}{99})\,(1 - \frac{4}{89})$$

$$= 0\!\cdot\!9800 \times 0\!\cdot\!9694 \times 0\!\cdot\!9579 \times 1\!\cdot\!154 \times 0\!\cdot\!9429 \times 0\!\cdot\!8989 \times 0\!\cdot\!9551$$

$$= 0\!\cdot\!85$$

Or, in the simpler form of Equation (1):

$$L_{\underset{w}{X\ J}} = 0\!\cdot\!910 \times 1\!\cdot\!154 \times 0\!\cdot\!810 = 0\!\cdot\!850$$

Writing the equation in these forms enables a ready comparison of the importance of various sources of loss; death rate of lambs is clearly more important in this case than mishap between joining and lambing.

The relative importance of the sources of differences in $L_{\underset{w}{X\ J}}$ between flocks can also be compared. Suppose we have another flock in which:

$$L_{\underset{w}{X\ J}} = 0\!\cdot\!850 \times 1\!\cdot\!020 \times 0\!\cdot\!800 = 0\!\cdot\!694$$

This flock differs from the one above, but in which component does it differ most? Using the percentage deviation technique described by Turner (1958) for wool weight and used by Turner and Dolling (1965) and Dun (1964) for reproduction rate, we have:

$$\% \text{ deviation in } L_{\underset{w}{X\ J}} = \% \text{ deviation in } E_{P_j} + \% \text{ deviation in } L_{BP} +$$

$$\% \text{ deviation in } L_{\underset{w}{X\ B}}$$

that is,

$$100\,(\frac{0\!\cdot\!850 - 0\!\cdot\!694}{0\!\cdot\!694}) = 100\,(\frac{0\!\cdot\!910 - 0\!\cdot\!850}{0\!\cdot\!850}) + 100\,(\frac{1\!\cdot\!154 - 1\!\cdot\!020}{1\!\cdot\!020})$$

$$+ 100\,(\frac{0\!\cdot\!810 - 0\!\cdot\!800}{0\!\cdot\!800})$$

giving $22 = 7 + 13 + 1$

The sums of the percentage deviations of the components will add approximately $(7 + 13 + 1 = 21)$ to the deviation in $L_{X\ J}$. From this we can see that a difference in $E_{PJ}$ contributes $\dfrac{7}{21} \times 100 = 33\%$ of the difference in $L_{X\ J}$ between groups, $L_{BP}$ contributes $62\%$ and $L_{X\ B}$ only $5\%$.

The total number of lambs of both sexes weaned is a useful commercial measure, since lambs are often sold at this age and it also marks the end of one stage of the lamb's life for scientific analysis of factors affecting reproduction rate. In considering the rate of replacement or expansion of a breeding flock, however, ewe lambs only are considered; and we come to the concept of 'net reproduction rate', that is, the number of ewe lambs which reach the age of first joining, produced by each ewe during her own lifetime in the breeding flock.

Consideration of lifetime production leads back to the influence of various factors on the terms of Equation (1). The most important of these factors is the age-structure of the flock, which depends on the ages of first and last joining of ewes (controlled by planning) and their rates of loss at each year of age while in the breeding flock (influenced by many factors). Rates of loss include deaths and withdrawals for other reasons, such as sale or additional culling; here only death rates will be considered.

Death rates are most conveniently estimated between one joining and the next, as the number of ewes dying during any interval, expressed as a percentage of the number joined at the beginning of the interval. The customary method of estimation is to follow a 'cohort' (ewes born in any one year) through life, and then to take mean values for a series of cohorts. Turner, Dolling and Sheaffe (1959) gave estimates of 1·5 to 2·6% for ewes aged $1\frac{1}{2}$ to $7\frac{1}{2}$ years, rising to 7·3% for ewes over $7\frac{1}{2}$ years. In a single drought year, these figures rose as high as 11·1% for some age-groups up to $7\frac{1}{2}$, and as high as 45·6 for some groups over $7\frac{1}{2}$.

Age-specific death-rates can be used to calculate the probability that an ewe entering the breeding flock will be alive at any subsequent time. Table 11.1 gives the probability of a ewe surviving (a) to the age of joining, $(1_X)$ and (b) to the age half-way between joinings, that is, the 'pivotal age' of the interval $(1_{\bar{X}})$. In each case, $1_{(X+1)}$ is calculated as $1_X(1 - d_X)$, and $1_{\bar{X}}$ as $1_X(1 - \dfrac{d_X}{2})$.

TABLE 11.1. Probability of survival for ewes entering the breeding flock, and age-specific rates of death, ewes lambing, total lambs born and weaned, and ewe lambs surviving to joining age.
(Data adapted from Turner & Dolling (1965))

| Age interval between joinings (years) | Pivotal age (years) | Deaths* ($d_x$) | Probability of survival to: | | Ewes lambing* ($E_{PJ}$) | Total lambs born* ($L_{BJ}$)‡ | Total lambs weaned* ($L_{x_w,J}$) | Ewe lambs reaching joining age* ($L_{EX,J}$) |
| --- | --- | --- | --- | --- | --- | --- | --- | --- |
| | | | Beginning of interval ($1_x$) | Pivotal age ($1_{\bar{x}}$)† | | | | |
| 1½–2½ | 2 | 0·024 | 1·000 | 0·988 | 0·819 | 0·842 | 0·621 | 0·292 |
| 2½–3½ | 3 | 0·026 | 0·976 | 0·963 | 0·864 | 0·907 | 0·723 | 0·340 |
| 3½–4½ | 4 | 0·015 | 0·951 | 0·944 | 0·876 | 0·958 | 0·773 | 0·363 |
| 4½–5½ | 5 | 0·021 | 0·937 | 0·927 | 0·917 | 1·052 | 0·855 | 0·402 |
| 5½–6½ | 6 | 0·026 | 0·917 | 0·905 | 0·917 | 1·098 | 0·891 | 0·419 |
| 6½–7½ | 7 | 0·026 | 0·893 | 0·881 | 0·915 | 1·112 | 0·870 | 0·409 |
| 7½–8½ | 8 | 0·073 | 0·870 | 0·838 | 0·902 | 1·102 | 0·824 | 0·387 |
| 8½–9½ | 9 | 0·073 | 0·806 | 0·776 | 0·881 | 1·065 | 0·783 | 0·368 |
| 9½–10½ | 10 | 0·100 | 0·747 | 0·710 | 0·865 | 1·043 | 0·707 | 0·332 |

* All rates are per ewe joined at beginning of interval.
† Corresponds to the statistic denoted $1_x$ by Andrewatha & Birch.
‡ Corresponds to twice the statistic denoted $m_x$ by Andrewatha & Birch (1954).

The $1_x$ and $1_{\bar{x}}$ columns of Table 11.1 mean, for example, that, of 100 ewes entering a breeding flock with these death rates, on the average 87 are still alive to be joined at $7\frac{1}{2}$ years of age, while 83·8 are alive half-way through the interval, that is, at lambing at 8 years.

The other columns give further age-specific information (Turner and Dolling, 1965) needed for calculating various aspects of reproduction rate, the $L_{BJ}$ column being $L_{PJ} \times L_{BP}$. The last column has been computed by applying a death rate of 6% between weaning and age at first joining, and assuming that half the lambs weaned are males. The latter assumption is not always correct; of 25,000 Merino births on 3 CSIRO stations over a period of 15 years the mean proportion of males has been 0·499, while survival rates are 2 to 5% lower in males than females (Lax and Turner, 1965). Another point of importance, concealed in the overall $L_{x\;J\atop w}$ figures of Table 11.1, is that the survival rate of twins is lower than that of singles; the average difference over all ages of ewe is 0·183, and there appears to be no consistent pattern with age of ewe (Turner and Dolling, 1965).

From Table 11.1, reproduction rates for flocks of varying age-structures can be calculated. A distinction needs to be drawn between two types of analysis:

(1) Estimation of productivity for a flock of fixed size,

(2) Estimation of the intrinsic rate of increase for a flock allowed to increase without restriction.

**Productivity of flocks of fixed size.** Table 11.2 shows the computations necessary to calculate the various components of reproduction rate for flocks of fixed size with varying age structures. If we have a flock of size N to which the date of Table 11.1 apply, then the number of ewes in each age-group will

be $\dfrac{1_{\bar{x}}}{\Sigma 1_{\bar{x}}}$N, where summation is from age 2 to the oldest age-group included.

The average number of lambs born per head will be $\dfrac{\Sigma 1_{\bar{x}} \cdot L_{BJ}}{\Sigma 1_{\bar{x}}}$, and so on

for other aspects of reproduction rate.

Tables such as Table 11.2 can be used:

1. To determine the age-structure which gives the highest reproduction rate for any given set of death and birth rates.

TABLE 11.2. Mean values of various aspects of reproduction rate for flocks of fixed size with differing age structures
(Ewes aged 2 to $\bar{x}$)
Basic data from Table 11.1

| Pivotal age ($\bar{x}$) | $\Sigma l_x$ | $\Sigma l_{\bar{x}}$ | Ewes lambing $\dfrac{\Sigma l_{\bar{x}} \times E_{PJ}}{\Sigma l_{\bar{x}}}$ | Total lambs born $\dfrac{\Sigma l_{\bar{x}} \times L_{BJ}}{\Sigma l_{\bar{x}}}$ | Total lambs weaned $\dfrac{\Sigma l_{\bar{x}} \times L_{x_W}J}{\Sigma l_{\bar{x}}}$ | Ewe lambs reaching joining age $\dfrac{\Sigma l_{\bar{x}} \times L_{EXj}J}{\Sigma l_{\bar{x}}}$ |
|---|---|---|---|---|---|---|
| 2 | 1·000 | 0·988 | 0·819 | 0·842 | 0·621 | 0·292 |
| 3 | 1·976 | 1·951 | 0·841 | 0·874 | 0·671 | 0·316 |
| 4 | 2·927 | 2·895 | 0·852 | 0·901 | 0·705 | 0·331 |
| 5 | 3·864 | 3·822 | 0·868 | 0·938 | 0·741 | 0·348 |
| 6 | 4·781 | 4·727 | 0·878 | 0·968 | 0·770 | 0·362 |
| 7 | 5·674 | 5·608 | 0·883 | 0·991 | 0·785 | 0·369 |
| 8 | 6·544 | 6·446 | 0·886 | 1·005 | 0·790 | 0·372 |
| 9 | 7·350 | 7·222 | 0·885 | 1·012 | 0·790 | 0·371 |
| 10 | 8·097 | 7·932 | 0·884 | 1·014 | 0·782 | 0·368 |

2. To calculate the number of surplus animals, at weaning or joining, for flocks of fixed size with a given age-structure. The number of ewes required for replacement is calculated as:

$$\frac{1 \cdot 000}{\Sigma 1_x} \times N$$

so the surplus ewes at joining are:

$$N \left\{ \frac{\Sigma 1_{\bar{x}} \times L_{EX\,j}}{\Sigma 1_{\bar{x}}} - \frac{1}{\Sigma 1_x} \right\}$$

The surplus males depend on the type of enterprise; in a flock buying its rams the surplus is:

$$\text{At weaning,} \quad \frac{\Sigma 1_{\bar{x}} \times L_{X_w\,J} \times \dfrac{N}{2}}{\Sigma 1_{\bar{x}}}$$

$$\text{At joining,} \quad \frac{\Sigma 1_{\bar{x}} \times L_{EXj} \times J}{\Sigma 1_{\bar{x}}} \times N$$

(assuming numbers of ewes and rams approximately equal). For a flock of 1000 ewes, the surplus ewes at joining would be (Table 11.2):

$$1000 \left(0 \cdot 348 - \frac{1}{3 \cdot 864}\right) = 89, \text{ if the oldest ewes were 5 and:}$$

$$1000 \left(0 \cdot 368 - \frac{1}{7 \cdot 932}\right) = 242, \text{ if the oldest ewes were 10.}$$

3. For 'standardizing' flocks with different age-structures. Obviously two breeding flocks cannot be compared directly if one ranges in age from 2 to 5 and another from 2 to 8. Comparison of the same age groups can be made from tables similar to 11.1 and 11.2.

**Intrinsic rate of increases.** The preceding sections have dealt with breeding flocks where the intake of young ewes is regulated so as to keep the flock at a constant size. But there are times when we want the flock to increase. How can we calculate its rates of increase? Andrewartha and Birch (1954), following Lotka (1925) and Fisher (1930), used a statistic $r_m$ which they called the 'innate capacity for increase', but which has also been called the 'intrinsic rate of increase'. If $N_o$ is the size of a breeding flock at one time, and $N_t$ its size after the elapse of time $t$, then we have:

$$\frac{N_t}{N_o} = e^r m^t$$

since the population increases logarithmically. In the previous section we introduced the concept of 'net reproduction rate', or the number of ewe lambs reaching joining age which each ewe produces in her lifetime. If we call this net reproduction rate $R_o$, then:

$$R_o = \frac{N_t}{N_o} = e^r m^t$$

where T is the length of a generation, that is, the average age at which the ewe produces the offspring which replace her. From this:

$$r_m = \frac{\log_e R_o}{T}$$

Values of $R_o$ for flocks with different age-structures (that is, for ewes with different lengths of life in the breeding flock) are shown in the third column of Table 11.3. The figures were of course used in calculating the last column

TABLE 11.3. Calculation of intrinsic rate of increase $(r_m)$ for flocks of different age-structures. increasing without restriction

(Ewes aged 2 to $\bar{x}$)

| Pivotal age in years $(\bar{x})$ | $\Sigma 1_{\bar{x}}$ | $\Sigma 1_{\bar{x}} \times L_{EX_{j.J}}$ ($R_o$) | $\dfrac{\Sigma 1_{\bar{x}} \times L_{EX_{j.J}} \times \bar{x}}{\Sigma 1_{\bar{x}} \times L_{EX_{j.J}}}$ (T) | $\dfrac{\log_e R_o}{T}$ $(r_m)$ | Multiplying factor (antilog$_e r_m$) |
|---|---|---|---|---|---|
| 2 | 0·988 | 0·288 | 2·000 | − 0.622 | 0·537 |
| 3 | 1·951 | 0·616 | 2·532 | − 0·192 | 0·825 |
| 4 | 2·895 | 0·958 | 3·057 | − 0·140 | 0·879 |
| 5 | 3·822 | 1·331 | 3·602 | +0·079 | 1·083 |
| 6 | 4·727 | 1·710 | 4·133 | +0·130 | 1·139 |
| 7 | 5·608 | 2·070 | 4·632 | +0·157 | 1·170 |
| 8 | 6·446 | 2·394 | 5·088 | +0·172 | 1·188 |
| 9 | 7·222 | 2·680 | 5·505 | +0·179 | 1·196 |
| 10 | 7·932 | 2·916 | 5·869 | +0·182 | 1·200 |

of Table 11.2, but here are required as totals and not as averages. The column means that if ewes are cast at $5\frac{1}{2}$ years, in a flock to which these birth and death rates apply, each ewe would contribute 1·331 females reaching age of first joining. If ewes were not cast till $10\frac{1}{2}$ years, the contribution would be 2·916.

The fourth column shows the calculation of T. The ewe cast at $5\frac{1}{2}$ has contributed 0·292 ewe lambs of joining age when 2 years old, 0·340 when 3 years, and so on; her average age at which these offspring were produced is thus

$$\frac{0·292 \times 0·988 \times 2 + 0·340 \times 0·963 \times 3 + \ldots\ldots\ldots\ldots}{0·292 \times 0·988 + 0·340 \times 0·963 + \ldots\ldots\ldots\ldots\ldots}$$

These values are tabulated in the fourth column of Table 11.3.

The fifth column gives values of the intrinsic rate of increase, $r_m$. This is on a yearly basis, since T was measured in years. If we want to find the factor by which the flock multiplies itself each year, we require $\text{antilog}_e r_m$, which is tabulated in the last column of Table 11.3.

These rates are theoretical and apply to a population which is, so to speak, in balance. If we have been running a flock at fixed size and then wish to build up, the intrinsic rate of increase will not apply for some years, as retention of all young stock will mean a sudden increase of entries over withdrawals, giving initially a higher rate of increase than indicated by the intrinsic rate. The intrinsic rate gives relative values which can be used for comparing various schemes and making long range forecasts. For example, if in the flock under discussion the death rates remained the same but the number of lambs born increased so that the $L_{EXj} \cdot _J$ values in Table 11.1 up to 6 years became 1·00, 1·20, 1·30, 1·40 and 1·50 respectively, the $r_m$ value would become 0·272 instead of 0·130, and the multiplying factor would be 1·313 instead of 1·139.

The figures in Table 11.3 apply to ewes. However, once a flock increasing without restriction has reached stability, the number of males will increase at the same rate, if we assume that all are retained and that sex numbers are equal.

### References

ANDREWARTHA H.G. & BIRCH L.C. (1954). *The Distribution and Abundance of Animals.* University of Chicago Press, Chicago.
DUN R.B. (1964). Skin folds and Merino breeding. 1. The net reproductive rates of flocks selected for and against skin fold. *Aust. J. Exp. Ag. Anim. Husb.* **4**, 376–385.

Fisher R.A. (1930 and 1958). *The Genetical Theory of Natural Selection*. Oxford University Press and Dover Publications, New York.

Lax J. & Turner H.N. (1965). The influence of various factors on survival rate to weaning of Merino lambs. 1. Sex, strain, location and age of ewe, for single-born lambs. *Aust. J. Ag. Res.* **16**, 981–95.

Lotka A.J. (1925). *Elements of Physical Biology*. Williams & Wilkins, Baltimore.

Turner H.N. & Dolling C.H.S. (1965). Vital statistics for an experimental flock of Merino sheep. II. The influence of age on reproductive performance. *Aust. J. Ag. Res.* **16**, 699–712.

Turner H.N., Dolling C.H.S. & Sheaffe P.H.G. (1959). Vital statistics for an experimental flock of Merino sheep. I. Death rates in relation to method of selection, age and sex. *Aust. J. Ag. Res.* **10**, 581–90.

# 12

# Nutrition

## PART I. INTRODUCTION

### R N B KAY

Any attempt to cover all aspects of the nutrition of herbivores in one short chapter would be so superficial that it would be of little value. Instead, therefore, this chapter deals with selected topics, the choice determined both by trying to guess the more important and rewarding subjects for study and also by the interests and experience of the contributors themselves. Running through many of the contributions is a desire to obtain quantitative information on the various adaptations of herbivores, tame and wild, to the inadequacies of their diets.

Consideration is first given to methods for studying the anatomy of the digestive tract, and for preservation of specimens for subsequent examination. This can well be made a field study of animals shot in their wild state. On the other hand, tame or captive animals are often needed for experiments on nutrition or the physiology of digestion that call for controlled conditions. Such experiments allow one to measure the nutrient requirements of livestock; these requirements and the formulation of rations are discussed next. Laboratory conditions are also required for microbiological studies of the micro-organisms of the digestive tract, and a mobile field laboratory is described in the following section. Other sections consider the utilization of urea by herbivores and its contribution to nitrogen economy, water metabolism and its study in the field, and the problems posed by toxic plants.

## PART II. TECHNIQUES AND EQUIPMENT FOR STUDYING THE ANATOMY OF THE DIGESTIVE TRACT

### R R HOFMANN

Methods of collecting and preserving specimens for studying the anatomy of the digestive tract have to be selected according to:
(a) the final objective of investigation
(b) the size of the animal (species, age)

(c) the state of the specimen at the time of obtaining it for preservation.

Initial preparations, equipment, chemicals and sequence of procedures are essentially different for collection of specimens for macroscopic and microscopic studies. They can, however, be combined where both types of examination are required.

## Macroscopic Preparations

The most satisfactory way to make the maximum use of the material available is to preserve the animal *in toto* via the blood vascular system. This, however, depends upon the size of the animal, the facilities available and the way the animal is obtained. In the field, animals up to 250 kg. (e.g. waterbuck), and in the laboratory, using a monorail hoist, animals up to 600 kg. (e.g. buffalo, eland, zebra) can be preserved by infusion of 8 to 10% formaldehyde solution under hydrostatic pressure into the arterial system (A. carotis comm.) so that the animal is fixed in a natural standing position (Paulli, 1909; Bevandic, 1960; Hofmann, 1965, 1966). This rules out the measurement of weights of single organs and reduces the opportunities for measuring capacities. It does, however, permit measurement of dimensions.

The great advantages include the speed of preservation (1—2 hours *post mortem*), saving of the whole body without deformation and maintenance of lifelike topographic relations *in situ*.

Animals can be used either after capturing, natural death or being shot in the neck (Hofmann, 1966).

Should equipment or facilities be insufficient for this method, animals may be preserved *in toto* in a lying position after upholstering the lower side of the body in order to prevent flattening and deformation.

**Techniques and equipment.** (Hofmann, 1966). Preservation of the digestive tract as *a whole* or in *single parts* after separation (exenteration) from the carcass can be carried out in different ways, again according to the size of the animal.

In any case, precautions have to be taken during field work, especially in warmer climates, against desiccation by sun and wind and heat decomposition due to delayed or time-consuming procedures (rubber groundsheets, wet covering cloths, spraying, protection from the sun). As the value of time-consuming weighing of single organs, especially glands, is subject to great doubts (greatly varying blood and secretion contents), preference may be

given, after taking dimension and capacity measurements, to immediate preservation via the regional vascular system (perfusion, infusion of fixative fluid under hydrostatic pressure or by injection).
Workable techniques are as follows.

(a) *Head and neck portion* (oral cavity, adherent glands, pharynx and oesophagus).

The quickest and safest way is to separate the entire head with a clean cut at about the 3rd cervical vertebra after ligatures have been applied to the carotids and jugulars on either side without skinning. A wooden plug is forced into the vertebral canal to compress the vertebral vessels. Initial *perfusion* of 8—10% formaldehyde solution through the carotid with one jugular vein open is followed by *infusion* of the fixative, the two jugulars and one carotid being closed. Hydrostatic pressure is sufficient at a height of 1·5 m. The volume of fixative fluid required is about equal to the volume of the head. The time needed for complete fixation is 1 to 2 hours. For a detailed description *see* Hofmann (1966).

Dissecting out of the cranial portions of the foregut in the fresh state requires skinning of the head and neck and has to be done in several phases but rules out vascular infusion. After careful removal of the big salivary glands (parotid, mandibular) two ventral cuts are made with a wide blade from the mandibular symphysis to the mandibular angle, close to the medial aspect of the bone on either side, loosening the tongue together with the sublingual gland and muscles. After exarticulation of the mandibular and hyoid joints, the mandibles are pulled down, the tongue is separated completely and the soft palate severed from the hard palate by a transverse cut. Combined blunt (roof of pharynx) and sharp (frenulum) separation will finally lead to exenteration of the tongue with accessory organs, pharynx, larynx and cranial portions of oesophagus and trachea together with accompanying structures. By splitting the neck muscles on the ventro-median aspect, exenteration may be continued down to the thoracic entrance.

Head and neck portions obtained separately in this way should be preserved immediately in a spacious container. In order to prevent distortion and deformation of the tongue, the latter should be pinned ventrally to a piece of timber or cork (to be floated upside down in the fixative). When these portions are taken out in the field, it is sufficient for a few hours to wrap them in a cloth soaked in 3% formaldehyde and keep in a polythene bag before transferring to a tank.

(b) *Thoracic, abdominal and pelvic portions* (oesophagus, stomach, liver, pancreas, small intestine, large intestine).

It is possible to take out the contents of the thoracic and abdominal cavities in one operation either together with the head and neck structures as described above or after cutting off the latter, cranial to the thoracic entrance. This method is satisfactory for an animal up to the size of an eland and has the advantage of allowing more elbow-space outside the body cavities. In larger animals (elephant, hippopotamus) ropes or chains attached to the thoracic aorta may be used in the technique as described below.

Provided the animal has been shot in the neck, the procedure is as follows: The animal is placed on a large rubber groundsheet in a recumbent position. The peritoneal cavity is opened paramedian to the *linea alba* cutting from the pelvic symphysis to the xiphoid cartilage with a blunt-pointed knife plus one transverse cut from the *linea alba* towards the *tuber coxae* on either side. Then the sternum is separated from the thorax thus opening the pleural cavities ventrally. The rib cartilages are cut; removal of the rib bones of one side may be necessary in large animals. The vascular trunks plus the oesophagus are cut off at the thoracic entrance and separated carefully from their dorsal attachments or branches to the vertebral column in cranio-caudal direction, pulling ventrally and caudally. The diaphragm is separated with a scalpel from the rib-cage close to its attachment.

Separation of the big vessels is continued backwards until the sacral end-division is reached, where the vessels are cut off. The pelvic symphysis is split ventrally to open the pelvic cavity after pushing the pelvic urethra (also the vagina) and the rectum as far dorsally as possible. Blunt separation of the pelvic structures will loosen their attachments, and those between the urogenital tract and the rectum, until finally the anus becomes separated from the skin by a circular incision. The exenterated viscera are displayed on the groundsheet.

After separation of the thoracic organs from oesophagus and disphragm and cutting the aorta several centimeters cranial to the *hiatus aorticus*, the aorta is split on its dorsal aspect and bulb-headed canulas are inserted and tied into the coeliac and the cranial and caudal mesenteric arteries. A ligature is laid to the hepatic portal vein, which vessel is subsequently severed close to the liver. The caudal vena cava is tied off cranial to the diaphragm before severing it from the heart thus leaving a closed vascular system which can be infused with formaldehyde.

Vascular infusion of the stomach should be supplemented (in ruminants, immediately) by infusion of 8—10% formaldehyde solution, using a funnel, via the oesophagus. This is done in smaller species by hand; in larger species with voluminous stomachs, an infusion system based on hydrostatic pressure is connected with the oesophagus. Similar infusion of the thin-walled intestinal tract via the anus has to be performed with great care and reduced pressure, as the gut is likely to burst. When working in the field, the organs have to be cooled and protected against desiccation. Finally, they are labelled, wrapped in wet cloths and placed in transport boxes and should be transferred for a few days to a fixation tank (8—10% formaldehyde or any other fixative). Where expense does not matter, other fixatives (e.g. Bouin's fluid) may be used for vascular injection particularly if histological examination is also intended.

*Equipment required.* At least one big knife, scalpels, blunt scissors, steel *post mortem* axe (sternum), saw (symphysis), saw-wire as for embryotomy (ribs of larger animals), haemostatic forceps, funnel, ligature string, bulb-headed canulas, Record syringe 50 cc., fluid container with handle and about 2 m. of rubber tubing, wet cloths, groundsheet, transport containers with lids (metal or plastic boxes), measuring cylinders, ropes.

This method permits the taking of measurements of the *stomach*, as smaller cut openings for emptying and measuring fluid capacity may be closed later by haemostatic forceps. It is, however, unsatisfactory for measuring the *intestine* without destroying the continuity of its vascular system. When measuring preserved guts, a shrinkage factor of at least 20% has to be taken into account. If measuring is the main objective, initial exenteration of the digestive tract can be carried out as described above; but the stomach should never be separated by cutting the oesophagus caudal to the diaphragm. The lesser omentum should be cut close to the liver. The greater omentum, especially in ruminants, should be cut close to its attachment to the stomach.

Adhesive attachment of the stomach to the abdominal wall (as in cattle, buffalo, etc.) requires blunt dissection. The intestine should be separated caudal to the pyloric sphincter. The spleen, if firmly attached to the stomach as in ruminants, can be separated carefully with a blunt instrument. It is easy to measure the capacity of the simple stomach but difficult to measure correctly that of the three or four compartments of the ruminants' stomach, particularly the omasum, without extensive damage. Removal of food before filling with fluid is best carried out after separation of omasum and abomasum from the recticulo-rumen.

Outside measurements of the stomach have to be taken along the curvatures from the cardia to the pyloric constriction and *vice versa* (note the state of contraction). Width measurements of fresh, empty stomachs are meaningless.

The *pancreas* requires great care when being separated from the abdominal wall and the intestinal tract. It should be separated from adjacent organs (diaphragm, stomach, liver, mesenteric plate, intestine) before any other organ.

The *liver* is firmly fused with the caudal vena cava and the diaphragm respectively. Separation from the latter requires careful dissection, as does also separation of other attachments in soliped animals. Separately preserved livers become flattened due to plasticity of their tissues unless upholstered.

Evaluation of liver-weights or even linear measurements are of doubtful significance due to functional phases and changing blood content.

Measuring the *intestine* longitudinally is done with a tape measure after separation from the mesentery along the mesenteric margin. Post-mortal contractions, especially of the surviving jejunal musculature, have to be taken into account. Width measures are best taken after splitting the tube at the antimesenteric margin. Longitudinal measures should always be related to body length (Ellenberger and Baum, 1943; Nickel, Schummer and Seiferle, 1960). Measuring the capacity of the various portions is again of doubtful value due to greatly varying states of contraction or distention in the time elapsing after death. In order to prevent bursting during attempts to fill the entire intestinal tube at once, double ligatures should be applied between its portions, at the duodeno-jejunal flexure, at the jejuno-ileac junction (end of ileac-caecal ligament), at the ileac-caecal junction, at the caeco-colic junction, at the colic-rectal junction, and at the recto-anal junction. The various portions may be cut apart between the ligatures as shorter portions are more easily dealt with. Cut openings can be closed by haemostatic forceps after infusion of fluid.

Weights of the opened gut should be taken after removal of mesenteries and reasonable drying of the exposed mucous membrane, especially after fluids have been used for measurement of capacity. Methods for weighing and measuring in the field have been described by Talbot and Talbot (1960, 1961), Talbot and McCulloch (1965), and Ledger (1963).

### Microscopic preparations

Speed is the most important factor to be observed in collecting histological specimens to prevent autolysis or heat-decomposition. Squeezing of soft tissues and cutting out unnecessarily large pieces has to be avoided. Bigger pieces of soft organs (pancreas, salivary glands, liver, etc.) must be cut down after 1–1½ hours of peripheral fixation. In turn, vascular injection of histological fixatives may be preferred, e.g. with guts of smaller animals. Such specimens, kept cool and moistened, have to be cut down to blocking size not later than 3 hours after infusion and transferred into the same fixative.

For details of preserving fluids and techniques, refer to recent textbooks on histological techniques (Romeis, 1948; Carleton & Drury, 1957; Haug, 1959; and Gurr, 1962). A few examples of fixatives are given below in brackets.

The sequence of collecting should be approximately as follows:

1. Pancreas (Bouin, Susa, Metzner, Launcy).
2. Glandular stomach (alcohol-formol, Carnoy).
3. Small and large intestines (formol, potassium bichromate-formol, Metzner).
4. Salivary glands (Helly, Bouin, Susa).
5. Liver (formol, Bouin, Helly).
6. Oesophagus, non-glandular stomach (formol, Bouin, Carnoy, Susa).

Fresh pieces of intestines are best preserved by cutting a short piece off the tube, tying one end and pouring the fixative into the lumen until it is reasonably filled. After tying off the upper end, the entire piece is placed into a container with the same fixative (Romeis, 1948).

Ingesta sticking to the mucosal wall will have to be sprayed off using a pipette filled either with the fixative or alcohol. In cleaning the papillated mucosa of the ruminant stomach, physiological saline has been used without adverse effect.

For collection in the field, 50 to 75 ml. plastic containers with lids may be used tightly fixed in wooden trays with handles (e.g. for 9 containers each). Provisional labelling is done with a grease pencil by code numbers on the lids.

It has to be remembered that taking specimens is time-consuming, and the collection of too many different samples from one animal may lead to delay and consequent decomposition.

*Equipment required.* At first essentially the same as for macroscopic work, with the addition of at least two fine-pointed forceps, fine scissors (straight and bent), fine scalpels, preparation needle, cork-plate, hedgehog-spikes, measuring cylinder, containers, fixatives.

**References**

BEVANDIC M. (1960). Schumer's formalinized horse as means for evident teaching. *Vet. Saraj.* **9**, 527–533.

CARLETON H.M. & DRURY R.A.B. (1957). *Histological Technique:* 3rd Edn. Oxford University Press. London.

ELLENBERGER W. & BAUM H. (1943). *Handbuch der vergleichenden Anatomic der Haustiere.* 18th Edn. Springer, Berlin.

GURR E. (1962). *Staining (animal tissues).* Leonard Hill, London.

HAUG H. (1959). *Leitfaden der mikroskopischen Technik.* Georg Thieme Verlag, Stuttgart.

HOFMANN R.R. (1965). Die Formalinfixierung ostafrikanischer Antilopen in stehender Korperhaltung (nach Paulli und Schummer). *Verh. anat. Ges. Jena* **115**, 547–550.

HOFMAN R.R. (1966). Field and laboratory methods for research into the anatomy of East African game animals. *E. Afr. Wildl. J.* **4** (in press).

LEDGER H.P. (1963). Animal husbandry research and wildlife in East Africa. *E. Afr. Wildl. J.* **1**, 18–29.

NICKEL R., SCHUMMER A. & SEIFERLE E. (1960). *Lehrbuch der Anatomie der Haustiere.* 2 Eingeweide. Paul Parey, Berlin.

PAULLI S. (1909). Formolinjektion zur Demonstration des Situs viscerum bei den Haussau-getieren. *Anat. Anz.* **34**, 369–375.

ROMEIS B. (1948). *Mikropischen Technik,* 15th Edn. Oldenbourg, Munich.

TALBOT L.M. & McCULLOCH J.S.G. (1965). Weight estimations for East African mammals from body measurements. *J. Wildl. Mgmt.* **29**, 84–89.

TALBOT L.M. & TALBOT M.H. (1960). Standard measurements of wild animals. Kenya Game Department, Nairobi. 2 p. (mimeographed).

TALBOT L.M. & TALBOT M.H. (1961). How much does it weigh? *Wild Life, Nairobi.* **3**, 47–48.

# PART III. RATIONS FOR MAINTENANCE AND PRODUCTION

R L REID

## Introduction

It is assumed that the research worker will be concerned with either (a) assessing the nutrient intake of animals in their natural environment or (b) feeding animals kept under some form of restraint for experimental purposes other than that of determining nutrient requirements. The deter-mination of nutrient requirements involves techniques which are outside the scope of this publication. The assessment of nutrient intake in a natural or semi-natural environment has been dealt with in a previous chapter. We are concerned here with the practical feeding of animals kept in captivity for various experimental purposes.

## Nutrient requirements

Many attempts have been made to summarize a very large volume of literature on nutrient requirements of ruminants. It is obviously impossible to present an adequate summary here. The most comprehensive publication is that produced recently by the Agricultural Research Council of Great Britain, *The Nutrient Requirements of Farm Livestock. No. 2. Ruminants: Technical Reviews* (1965).* The relevant tabular information is summarized in a second publication, *The Nutrient Requirements of Livestock. No. 2. Ruminants: Summaries of Estimated Requirements* (1965). These publications deal only with sheep and cattle, under the following main headings (1) Water, (2) Major mineral elements, (3) Minor and trace elements, (4) Vitamins, (5) Protein, (6) Energy, and (7) The expression of nutrient requirements as dietary concentrations. The aim was to estimate the requirements of cattle and sheep for each nutrient for maintenance, growth, pregnancy, lactation, and for sheep, wool growth. The authors point out that they considered primarily the needs of ruminants under conditions that prevail in the United Kingdom, and that some of the figures put forward would almost certainly have to be modified for countries with very different climatic conditions and in which different breeds and feeds and different systems of management exist. In the virtual absense of precise data on other species which are likely to be studied in IBP research projects, this publication at least forms a basis from which to extrapolate to other ruminant species. There seems no reason to believe that the basic nutritional principles governing nutrient requirements will differ so markedly as to make some intelligent extrapolation invalid.

(a) **Energy requirement.** The first requirement which must be met is that for energy. Energy was discussed in Sections 2 and 3; the remarks here will supplement those. The above publication suggests that requirements should be expressed in terms of metabolizable energy (M.E.) which is that part of the digestible energy of the ration which is not excreted in the urine or lost as

*Obtainable from H.M.S.O., 49 High Holborn, London, W.C.1 (30s. net).

methane from fermentation in the rumen. Methods of calculation of M.E. range from refined estimates, with comparatively low errors, which involve a detailed knowledge of the energy values of the feeds used and of the requirements to be met, to simplified calculations for use when adequate information, particularly on the energy value of feedstuffs and feed mixtures, is not available. Unfortunately, although there is a very large amount of data on energy requirements for almost all forms and levels of production in cattle and sheep, there is a marked scarcity of accurate data on the energy values of feedstuffs and feed mixtures which may be used to meet these requirements. This is because the most accurate method of determining the M.E. value of a feed, that is, the direct determination of the heat of combustion of the feed, faeces and urine and the amount of methane produced by the animal, is tedious and requires facilities which will not usually be available. The amount of information available is therefore small and is largely confined to dried herbage and hays, mostly from temperate grass species. Few data are available on the grain and protein concentrates which may be readily available in other temperature or in tropical areas.

The energy content of the dry matter of most feeds and faeces (other than when feeds of a high oil content are given) is fairly constant at 4·4 kcal/g dry matter. It is relatively easy to determine dry matter (D.M.) digestibility either directly in a feeding experiment, or by an *in vitro* technique (Alexander and McGowan, 1961; Tilley and Terry, 1963) where it is known that this is valid. Apparently digested energy (D.E.) can thus be calculated. Alternatively, the digestible energy content of the ration (kcal/g D.M.) can be predicted with a high order of accuracy from D.M. digestibility by means of the equation (Moir, 1961):

$$y = 0·0462x - 0·158$$
where $y$ = digestible energy content of the ration
and   $x$ = D.M. digestibility

This equation was derived from an examination of published data involving many different foodstuffs and levels of intake for both cattle and sheep over the range 30—83% D.M. digestibility.

Since losses of urine and methane energy are reasonably constant at 18% of D.E., the M.E. content of the feedstuff or diet can be calculated with reasonable accuracy simply by multiplying D.E. by 0·82.

In the absence of D.M. digestibility data, the best approximation is to calculate M.E. directly from published tables of total digestible nutrient

TABLE 12.1. *Energy Requirements for Maintenance of Adult Ruminants* (from the Nutrient Requirements of Livestock. No. 2. Ruminants: Summaries of Estimated Requirements, 1965)

| Energy concentration of diet (kcal M.E./ g D.M.) | 1·8 | | | | | 2·6 | | | | |
|---|---|---|---|---|---|---|---|---|---|---|
| Requirements | kcal M.E. per kg$^{0.73}$ | kcal M.E. per kg | kcal D.E.* per kg | g D.D.M.† per kg | kg diet per day (D.M.) | kcal M.E. per kg$^{0.73}$ | kcal M.E. per kg | kcal D.E.* per kg | g D.D.M.† per kg | kg diet per day (D.M.) |
| Cattle | | | | | | | | | | |
| 360 kg | 132 | 26·7 | 32·5 | 7·5 | 5·4 | 122 | 24·8 | 30·2 | 6·9 | 3·5 |
| 500 kg | 132 | 24·6 | 30·0 | 6·8 | 6·9 | 122 | 22·8 | 27·8 | 6·3 | 4·4 |
| 590 kg | 132 | 23·7 | 29·0 | 6·6 | 7·8 | 122 | 21·8 | 26·6 | 6·1 | 5·0 |
| Sheep | | | | | | | | | | |
| 70 kg | 91 | 28·9 | 35·2 | 8·0 | 1·1 | 85 | 26·7 | 32 5 | 7·4 | 0·7 |

$$* \text{ D.E.} = \frac{\text{M.E.} \times 100}{82}$$

† On assumption that gross energy of dry matter of diet is 4·4 kcal/g.

content (TDN) by using the factor 3·6, the average M.E. per $g$ TDN calculated from Swift's (1957) value of 4·4 kcal D.E. per $g$ TDN.

The M.E. of diets with a high energy content is used for maintenance with a higher efficiency than that of diets with a low energy content (roughage diets). However, if it is assumed that diets fed only for maintenance are unlikely to contain less than 50% roughage, the error introduced by neglecting this systematic variation will be small.

The energy requirements of adult ruminants are a function of metabolic body size, rather than of body weight (W), and the unit of metabolic body size commonly accepted is $W^{0·73}$. Requirements per unit of body weight thus decrease as body weight increases. However, the variation in requirements per unit of body weight within a given species is not so great as to lead to substantial error if a single value is used when accurate estimates are not required, especially as variation between individuals tends to be of the order of $\pm 10\%$. These points are illustrated by the data in Table 12.1. Two diets

TABLE 12.2. Nutrient requirements for maintenance of adult ruminants (from The Nutrient Requirements of Livestock. No. 2. Ruminants: Summaries of Estimated Requirements, 1965)

| Requirement | Energy concentration of diet (kcal M.E./g D.M.) | |
|---|---|---|
| | 1·8 | 2·6 |
| Adult cattle (500 kg) | | |
| D.M. (kg/day) | 6·9 | 4·4 |
| Concentration in D.M. (%) | | |
| Dig. crude protein | 3·1 | 4·1 |
| Calcium | 0·3 | 0·45 |
| Phosphorus | 0·4 | 0·60 |
| Adult sheep (70 kg) | | |
| D.M. (kg/day) | 1·1 | 0·7 |
| Concentration in D.M. (%) | | |
| Dig. crude protein | 4·3 | 5·8 |
| Calcium | 0·55 | 0·90 |
| Phosphorus | 0·45 | 0·75 |

are considered. The first, with an M.E. content of 1·8 kcal/g, is equivalent to a roughage of 'fair quality'; the second, with an M.E. content of 2·6 kcal/g, is equivalent to a 1:1 mixture of good quality hay and grain. This table also expresses energy requirements in terms of D.E. and digestible dry matter (D.D.M.) per kg body weight.

In general, it will be seen that the energy requirements for maintenance of adult ruminants can be met by a roughage diet (containing at least 6—8% crude protein—*see* Table 12.2) fed at a level of 7—9 *g* digestible D.M. (D.D.M.) per kg body weight. Such a ration should have a D.M. digestibility of 55—60%, so that the required intake of D.M. is about 15 *g* per kg. This intake is likely to be 60—75% of *ad libitum*.

It will be obvious that the requirement for a diet of high energy value is considerably less and may be as low as 30% of *ad libitum* intake. Feed costs may be reduced considerably, but animals fed at such low levels tend to develop peculiar habits. Sheep will chew their own or other sheep's wool and both sheep and cattle may avidly consume sawdust, straw bedding, etc.

There is one additional factor whose relative importance may be difficult to assess. The requirements discussed above are those of domesticated species of ruminants. Wild species will probably suffer a variable degree of psychological 'stress' when kept under restraint, and this will lead to substantial increases in energy requirements which may persist for a considerable period of time. Unfortunately, the presence of a stress response may be associated with a degree of inappetence. The only practical suggestion is that feeding should be *ad libitum* during the initial 'settling-down' period.

No attempt can be made here to summarize data on energy requirements for growth or lactation. The efficiency with which M.E. is used for fattening varies greatly with the energy concentration of the diet, and requirements can only be calculated on the basis of the level of production (rate of body weight gain, milk production etc.) desired or expected. Appropriate summaries are given in The Nutrient Requirements of Livestock (1965).

(b) **Protein requirement.** None of the methods used to estimate protein requirement are free from errors, which might account for differences between estimates of minimum requirement obtained by a factorial method and estimates obtained from feeding experiments. Recommended levels of feeding are often higher than estimated requirements, and, for practical purposes, it may be desirable to exceed the recommended minimum. It is often forgotten that protein is quite as good a source of energy as carbohydrate, though usually more expensive. Some data for maintenance of adult sheep and cattle are included in Table 12.2.

There may, however, be situations in which diets provided for captive animals must be based largely on roughages of low quality and protein content. At levels of crude protein below about 6%, voluntary consumption of roughages declines markedly as protein content declines. A point is reached at which voluntary consumption falls below the level necessary to meet energy requirements. The addition of small quantities of a legume hay such as lucerne hay will usually restore intake to an adequate level. Small quantities of grain and protein concentrate, and sometimes protein concentrate alone (such as linseed meal) are also effective. No firmer recommendations can be made, however, particularly as to relative quantities of low-quality roughage and supplement, because these depend on characteristics of the low-quality roughage which are inadequately defined.

### Practical aspects of feeding

**Pregnancy.** At least three reasons can be given for emphasizing the importance of nutrition in pregnancy, particularly in late pregnancy. First, undernourishment (energy deficiency) in late pregnancy prejudices foetal growth and development, suivival of the newborn young and subsequent milk production by the dam. Secondly, and in spite of the above, there has been no serious attempt until recently to estimate energy requirements at the several stages of pregnancy. Estimates based on recent work have been made for cattle in The Nutrient Requirements of Livestock (1965) but no data are given for sheep, because of lack of experimental infoi mation. Thirdly, it is the author's experience that iations offered to ruminants in late pregnancy are often inadequate. This is probably because published feeding standards for cattle allow an energy increase of approximately 75% in late pregnancy, which includes a so-called 'steaming-up' allowance. This latter term carries a luxury connotation unless high levels of milk production are required. In fact, the minimum requirements for pregnant cows, estimated by a factorial method, are remarkably similar to previous recommendations.

More precise techniques for controlling nutrition at a maintenance level in late pregnancy are now available (Reid and Hinks, 1962), and these have been used to estimate nutrient requirements of pregnant ewes. Direct estimates in terms of metabolizable energy, using calorimetric methods, have not yet been made; and the figures were obtained in terms of digestible organic matter (D.O.M.). For Scottish Blackface ewes, the additional DOM requirement is 100 g per kg foetus (Russel and Reid, unpublished data). This compares with a previous (now recalculated) estimate of 80 g per kg obtained in earlier experiments with Merino Sheep (Reid, 1963). Both values were obtained with predominantly roughage diets. The metabolizable energy requirement can be calculated as approximately 350 kcal per kg foetus.

**Standardization of diets.** Far too little attention has been paid to this point. Many experiments are difficult to compare because of wide differences in the type of diet used. The varying availability and price of feedstuffs makes standardization between different research units difficult to achieve. Standardization is easiest when the number of different feeds in the ration is fewest. It is most difficult when grass or grass-clover hays form the bulk of the roughage component and when this constitutes the major part of the ration. The wide variations in nutritive value of such hays can be largely overcome in warm temperate areas if lucerne hay forms at least 50% of the roughage component. In these circumstances, the nutritive value of the other roughage is of considerably less significance.

**Formulation of diets.** There is no reason to believe that complex diets, containing up to six different constituents (or even more) have any significant nutritional virtues. Over the past 15 years, the author has used experimental diets with only two constituents in many experiments in which the study of nitritional differences between diets was not an objective. No problems or deficiencies were encountered in several hundred experimental sheep fed indoors for periods as long as two years. The diets were (1) 1:1 chopped lucerne and wheaten hays and (2) 1:1 chopped lucerne hay and oat grain. No special attention was paid to sources of supply, yet, under Australian warm temperate conditions, there was a remarkable consistency in digestibility and chemical composition whenever these were measured. There seems no reason why wheaten hay (6% C.P.), which would be unavailable in many areas, should not be replaced by hays of lower quality, particularly if only maintenance requirements need to be met. Likewise, replacement of oat with maize grain, which may be more readily available, would give a ration adequate for all except very high levels of, e.g. milk production. *Ad libitum* intakes of these diets by non-lactating sheep average 25—30 g per kg body weight.

The importance of lucerne hay, in proportions as low as 25% of a ration, is emphasized because its high content of calcium and cobalt correct the serious deficiency of calcium and possibly cobalt in grains and the possible deficiency of cobalt in hays. Although its carotene content is often adequate, in contrast to grass hays, it is desirable in the author's opinion and experience to take other measures to ensure that deficiencies of vitamins A and D do not occur in hand-fed animals, particularly when housed continuously indoors. This can be most easily achieved by administration of massive doses of these vitamins at three-monthly intervals. Oral administration of a suitably-diluted water-dispersible vitamin A concentrate, in a single dose of 25,000 I.U. per kg body weight, is a convenient method. Subcutaneous administration of a concentrated calciferol preparation, in a single dose of 20,000 I.U. per kg, is the preferred method of vitamin D administration. Over-dosage can be dangerous.

The signs of various vitamin and mineral deficiencies in ruminants are adequately described in standard texts on animal nutrition (e.g. Underwood, 1962), and no attempt will be made to describe them here. Vitamin C is not required in the diets of most animals, and most vitamins B are synthesised microbially in the rumen. In general, it can be suggested that the main hazards are deficiencies of vitamins A and D and of calcium (in predominantly grain rations), phosphorus (in some roughages) and possibly cobalt (certain areas only, especially in grain rations). It also seems reasonable to suggest that supplementation of diets with these particular vitamins and minerals, and in tropical areas with sodium chloride, may be the simplest method of ensuring that the maintenance of animals being completely hand-fed is not complicated by such deficiencies.

Table 12.3 presents data on the digestible crude protein (D.C.P.), metabolizable energy, calcium and phosphorus contents of a selection of feedstuffs likely to be used in rations when animals are completely hand-fed. It must be emphasized that these are average or ranges of values obtained from data in the various publications referred to in this Section and that they may be inaccurate for the particular conditions or environment. This is especially so for roughages. Calcium and phosphorus contents of supplements commonly used to supply additional amounts of these minerals are also given in this Table.

The series of publications by the National Research Council of the U.S. National Academy of Sciences entitled *Nutrient Requirements of Domestic Animals* contains much relevant nutritional information which supplements that given briefly here. In particular, signs of dietary deficiencies are discussed and tables of feedstuff composition are presented. Nos. 3, 4 and 5 in this series deal respectively with dairy cattle, beef cattle, and sheep. These

TABLE 12.3. Nutrient content of feedstuffs.

| Feedstuff | Dry matter<br>% | Digestible crude protein<br>% D.M. | Metabo-lizable energy kcal/g D.M. | Calcium<br>% D.M. | Phos-phorus<br>% D.M. |
|---|---|---|---|---|---|
| 1. Clover and timothy hay (30–50% clover) | 90 | 4–6 | 1·8–2·1 | 0·6–0·8 | 0·15 |
| 2. Cowpea hay | 90 | 11·0 | 2·0 | 1·20 | 0·20 |
| 3. Grass hay (early cut) | 90 | 5–6 | 1·8–2·2 | 0·40 | 0·30 |
| 4. Grass hay (late cut) | 90 | 3–4 | 1·6–1·8 | 0·40 | 0·20 |
| 5. Lucerne (alfalfa hay) | 90 | 10–12 | 1·9–2·1 | 1·2–1·8 | 0·20 |
| 6. Maize (corn) cobs (ground) | 90 | 0·0 | 1·6 | 0·10 | 0·05 |
| 7. Maize (corn) and cob meal | 85 | 6·0 | 2·9 | 0·04 | 0·25 |
| 8. Prairie hay (mature) | 90 | 1·0 | 1·6–1·8 | 0·30 | 0·10 |
| 9. Soybean hay | 90 | 10·0 | 2·0 | 1·10 | 0·20 |
| 10. Oat or barley straw | 90 | 0·8 | 1·6 | 0·25 | 0·10 |
| 11. Wheat straw | 90 | 0·5 | 1·5 | 0·20 | 0·10 |
| 12. Maize, sorghum or millet silage | 25–30 | 4–6 | 2·2–2·4 | 0·30 | 0·30 |
| 13. Barley (grain) | 90 | 7–9 | 3·1 | 0·05 | 0·35 |
| 14. Maize (corn) (grain) | 90 | 7–9 | 3·3 | 0·02 | 0·30 |
| 15. Oats (grain) | 90 | 7–9 | 2·8 | 0·05 | 0·30 |
| 16. Sorghum (grain) | 90 | 7–9 | 3·1 | 0·04 | 0·30 |
| 17. Wheat (grain) | 90 | 7–9 | 3·1 | 0·04 | 0·35 |
| 18· Cottonseed oil meal | 90 | 33 | 3·2 | 0·20 | 1·20 |
| 19. Linseed oil meal | 90 | 26 | 2·9 | 0·40 | 0·75 |
| 20. Palm kernel meal | 90 | 17 | 2·7 | 0·20 | 0·50 |
| 21. Peanut (groundnut) oil meal | 90 | 37 | 3·3 | 0·15 | 0·60 |
| 22. Safflower oil meal | 90 | 17 | 2·0 | 0·25 | 0·70 |
| 23. Molasses (cane) | 75 | 5·0 | 2·6 | 1·0 | 0·10 |
| 24. Bone meal (steamed) | | | | 30·0 | 14·0 |
| 25. Dicalcium phosphate | | | | 26·5 | 20·5 |
| 26. Disodium phosphate | | | | Nil | 8·5 |
| 27. Limestone (ground to pass 120 mesh sieve) | | | | 39·0 | Nil |

publications are available from the Publication Office, National Academy of Sciences—National Research Council, 2101 Constitution Avenue, Washington 25, D.C., U.S.A.

The most comprehensive publication on feedstuff composition is 'Feeds of the World, Their Digestibility and Composition', by B. H. Schneider (West Virginia Agricultural Experiment Station, Morgantown, West Virginia, U.S.A., 1947) copies of which are still available.

### References

ALEXANDER R.H. & McGOWAN M. (1961). A filtration procedure for the *in vitro* determination of digestibility of herbage. *J. Br. Grassld. Soc.* **16**, 275–277.

MOIR R.J. (1961). A note on the relationship between the digestible dry matter and the digestible energy content of ruminant diets. *Aust. J. Exp. Ag. Anim. Husb.* **1**, 24–26.

REID R.L. (1963). The nutritional physiology of the pregnant ewe. *J. Aust. Inst. Ag. Sci.* **29**, 215.

REID R.L. & HINKS N.T. (1962). Studies on the carbohydrate metabolism of sheep. *Aust. J. Ag. Res.* **13**, 1092–1111.

SWIFT R.W. (1957). The caloric value of TDN. *J. Anim. Sci.* **16**, 753–756.

TILLEY J.M.A. & TERRY R.A. (1963). A two-stage technique for the *in vitro* digestion of forage crops. *J. Br. Grassld. Soc.* **18**, 104–111.

UNDERWOOD E.J. (1962). *Trace Elements in Human and Animal Nutrition*. Academic Press, Inc., New York & London.

## PART IV. MICROBIOLOGY OF THE DIGESTIVE TRACT

### P N HOBSON and S O MANN

All herbivorous animals depend on microbial activity in some portion of the gut for digestion of many of the constituents of herbage. In some cases the microbial digestion precedes digestion of the food by enzymic secretions from the animal body ('gastric' digestion), and in other cases microbial digestion comes after this. The end-products of food digestion which are available to the animal can thus vary in quantity and quality. However, even among animals with a similar gut anatomy the population of food-digesting microorganisms varies with the food eaten and may vary with the host species. The ability of an animal to thrive on a particular terrain, then, may depend not only on such things as its eating habits but also on its ability to support a gut microflora capable of digesting the herbage present. While microbiology may play no part in assessing the actual productivity of herbivorous animals, this being done by weighing and measuring techniques, it is possible that analysis of the gut flora may point to a reason why one animal may thrive better than another. For instance, a primary digestion of the feed by micro-organisms might detoxify an otherwise toxic herbage before it reached the sites of absorption in the small intestine. Microbiological studies are thus necessary adjuncts to ecological studies in a complete assessment of the productivity of wild herbivores.

As mentioned above, the part played in food digestion by gut micro-organisms depends on the relative positions of the microbial and gastric digestions and whether the products of microbial digestion can be fully absorbed into the host body. Dissection of the animal, tying off of the various portions of the gut, and weighing them full and empty of digesta can be carried out in the field and, combined with measurements of microbial and fermentation product concentrations, will give an idea of the possible contribution of microbial digestion to the total. Further information can be obtained from rates of passage of digesta through the gut as the extent of microbial digestion of food depends on this as well as on the rates of microbial fermentation. However, there seems to be no method of determining rates of passage of food through the alimentary tract of completely wild animals. Methods for use on tame or partially restricted animals are dealt with elsewhere.

It may be possible in the case of some wild ruminants to catch and hold them (perhaps after tranquilization) and obtain rumen samples by stomach tube. However, the texture of the rumen ingesta or the anatomy of the animal may preclude this in most cases, and the animal will have to be killed for dissection. Dissection is, in any case, needed to sample organs other than the rumen. For optimum determination of fermentation rates and viable bacteria, not more than half an hour should elapse between the killing of the animal and the setting up of experiments. Because of variations in the feeding cycle the measurements described here need to be carried out at different times of the day for optimum comparisons between different animal species. The following measurements assume that laboratory facilities are near the site of the 'kill'.

## Microbiological Work

After opening the abdominal cavity, portions of the gut can be tied off, either *in situ* or after removal of the alimentary tract from the body cavity.

If a rumen sample only is needed, this can be taken without removal of the organ from the animal. A large incision is made and the rumen contents mixed by hand. pH is determined by insertion into the contents of a combined calomel-glass electrode from a portable pH meter. A suitable large volume (say 200 ml) of digesta can then be transferred to a container and taken to the laboratory. The contents of large organs such as the caecum may also be mixed and the pH determined *in situ*. The contents of small organs have to be gently expressed into a suitable container. If there is some distance to travel to the laboratory, the samples should be transported in a thermos flask or in small bottles kept in water at body temperature (although about 20°—23° C. has been suggested as a better temperature for keeping rumen contents). Unless the containers are completely filled with the sample, the air in them should be displaced with $CO_2$.

In some cases the digesta may be so liquid that samples may be measured by volume. A 10 ml graduated pipette with the jet completely cut off to give an orifice of the diameter of the pipette tube can be used for measuring small samples. If the digesta are rather solid, then samples must be weighed out.

One gram (or 1 ml) of samples of different rumen contents have been found to give reproducible counts of bacteria, but if the digesta contains large pieces of herbage it may be necessary to make the first dilution from a greater mass (up to 10 g) of digesta. The samples are diluted and media inoculated as soon as possible after removal from the animal, but inoculated cultures will keep for some hours at ambient temperature until they can be incubated at the body temperature of the animal sampled.

*In vitro* fermentation rates and a measure of the growth rate of the rumen microorganisms can be determined from samples incubated in a closed flask with a hypodermic syringe manometer (El-Shazly and Hungate, 1965). This method may more conveniently be used with 10 g rumen sample, 20 ml bicarbonate-salts solution and 0·5 g herbage substrate in a 50 ml conical flask. Substrates can be finely-milled fresh herbage, or, to obtain a uniform substrate which can be kept, the herbage can be cut up, freeze-dried, and then milled. This method gives a short-term fermentation rate. For determination of, say, cellulose digestion, a longer-term incubation could be used with a final chemical estimation of cellulose consumed, but this might not be possible in the field.

For microscopic work, samples may be examined immediately—either 'live' or in fixed, stained, films. For later examination, samples can be fixed by addition of an equal volume of 10% formalin in water. Microscopic examination of this can be used for identification of protozoa, for phase-contrast examination of bacteria, for testing of iodophilic bacteria and for making stained films for morphological identification of bacteria (although there is much pleomorphism within rumen bacterial species). These samples could also be used for other tests, for instance, serological identification of bacteria with fluorescent antibodies (Hobson *et al.*, 1962). After suitable dilution the formalin-fixed sample can be used for counting protozoa (a simple method similar to that of Purser and Moir, 1959, can be used) and for counting bacteria either microscopically by a chamber of fixed-film method, or by an electronic counter (Hobson *et al.*, 1966).

## Biochemical analysis

Biochemical analyses must generally be done with the equipment available in a large laboratory, so that samples need only be fixed in the field for later analysis. However, to ensure that the concentrations measured are similar to those in the rumen, speed in fixing after taking the samples is essential. Only some general methods are given here; for particular estimations it may be better to fix the sample by using the first stage of the determination procedure. Centrifugation at low speeds may be carried out in the field. Samples for volatile fatty acid (VFA) analyses may be fixed with an equal volume of 10N $H_2SO_4$ saturated with $MgSO_4$, or 4 volumes with 1 volume of 5N $H_2SO_4$ containing 25% metaphosphoric acid. After centrifuging, this latter may be used directly for VFA analysis on some forms of gas-liquid chromatograph, although we have found that an interfering substance in hill-sheep rumen fluid makes it necessary to distill off the VFA before analysis. Steam distillation of the acid samples can be used to determine total VFA. Lactic and succinic acids can also be determined from acid-fixed samples. Samples for $NH_3$ determination can be fixed in an equal volume of 0·1 N HCl for later determination of ammonia by a micro-diffusion method. Determination of protein and non-protein nitrogen may be made on samples acidified to a standard pH. Samples acidified to a standard pH may also be used for later determination of metals, but in both cases some preliminary experiments may be needed to find the correct conditions for fixing. Samples for determination of soluble carbohydrates can be fixed with $HgCl_2$ (0·1%) or by acidification; these later may be used for other carbohydrate determinations.

Samples for later lipid analysis can be fixed in 2 volumes of methanol. Formalin-fixed samples also may be used for herbage analysis.

The best general reference text for descriptions of the rumen bacteria and protozoa is by Hungate (1966). From this, reference may be made to the original papers. This book also describes methods and some media used in the culture of rumen bacteria. There exist no simple schemes for classification of most of the anaerobic, feed-digesting, rumen and intestinal bacteria. A large number of properties must be determined and comparison made with those described in texts such as Hungate's (1966) or Bryant's (1959) or the original papers. Smith (1965*a*, 1965*b* and other papers) gives details of isolation and counting of many common intestinal bacteria and gives results for many kinds of animals. Mushin and Ashburner (1964) give details of coli-aerogenes isolates from various animals and birds, and Barnes (1964) gives the properties of various types of streptococci from the intestines of animals. 'Identification methods for microbiologists' (1966) gives notes on classification of Bacteroidaceae and lactic acid bacteria amongst others. Yeasts are not usually of importance in the gut flora of herbivores, but descriptions of some isolated have been given (Van Uden *et al.*, 1958; Clarke and DiMenna, 1961).

**Notes on some media for alimentary microorganisms**

Protozoa can only be studied microscopically, as can some uncultured rumen bacteria. Some of these latter are large and morphologically distinguishable, but the percentage of uncultured forms in the small rumen bacteria is unknown. Viable counts are generally only a small percentage of total counts. Most of the important food-digesting rumen bacteria must be cultured under the special anaerobic, high-$CO_2$-concentration, conditions described by Hungate (1966). This involves $CO_2$ freed from oxygen either *in situ* or as specially-purified cylinder $CO_2$. The papers cited in the previous paragraph give details of some media, and the media to be described have either been designed especially for growth of rumen bacteria (and need to be prepared and incubated under $CO_2$) or were originally used for culturing bacteria from other materials and have subsequently proved reliable for growth of certain rumen organisms. *Medium* 8 (*M*8.) The composition of this medium is such that the majority of rumen bacteria should be capable of growth. *Composition.* Rumen fluid (centrifuged) 20 ml, Bacto-casitone 1·0 g, yeast extract 0·25 g, Minerals 1 and 2 (Hobson and Mann, 1961) 15 ml each, sodium lactate (70% w/w) 1·0 ml, resazurin (0·1% w/v) 0·1 ml, distilled water 39 ml, agar 2·0 g, solution A 10 ml. The above excluding solution A, sterilized at 15 lb/15 min. Solution A (sterilized by filtration and added to the remainder of the sterilized medium) glucose 0·2 g, cellobiose 0·2 g, maltose 0·2 g, NaHCO$_3$ 0·4 g, cysteine HCl 0·05 g, distilled water 10 ml. *Starch medium.* This medium was originally compiled for the growth of starch-fermenting bacteria but appears to act as a non-selective medium and will in most cases give higher viable counts than those obtained with M8 at a 3-day incubation period. *Composition. Base:* rumen fluid (centrifuged) 20 ml, Bacto-casitone 1·0 g, yeast extract 0·25 g, Minerals 1 and 2 (Hobson and Mann, 1961) 15 ml each, resazurin (0·1% w/v) 0·1 ml, agar 2·0 g, distilled water 25 ml. Sterilized 15 lb/15 min. *Starch solution* 5 g soluble starch in 150 ml distilled water heated in boiling water bath until solution 'clears', bottled in 15 ml amounts and sterilized 5 lb/30 mins. *Solution A* (sterilized by filtration) cysteine HCl 0·05 g, NaHCO$_3$ 0·4 g, distilled water 10 ml. On removing base from autoclave add solution A (10 ml) and starch solution (15 ml). Mix and tube. We have used diluting fluids similar in composition to M8, with Tween 80 (0·1%) but omitting the carbohydrates. Dilutions are made under CO$_2$. Media similar in composition to M8 have been successfully used by a number of workers (for review *see* Hungate, 1966). All culturable rumen or intestinal bacteria should grow in these media, but in some cases counts of specific types may be required.

Methods and media for the culturing of bacteria with a specific function are given in the following references: Cellolytic bacteria (Hungate, 1966); Lipolytic bacteria (Hobson and Mann, 1961); Proteclytic bacteria (Blackburn and Hobson, 1962); Hemicellulose-fermenting (Hobson and Purdom, 1961); Selective and differential media: Lactobacilli (Rogosa *et al.*, 1951) (de Mann *et al.*, 1960); *Veillonella* and Peptostreptococci (Rogosa, 1964); *Streptococcus bovis* (Barnes, 1956; Madrck and Barnes, 1958 and 1962). For many purposes total viable counts or identification to generic level are all that is required, and further identification of isolates would involve the use of other media to determine specific properties. Details may be found in the papers cited above or in the previous paragraph. Gram films made from colonies growing in the anaerobic media (M8 and starch) may show the proportions of different morphological types (and also give some indication of identity) present in the original samples, provided that a large number of colonies from the higher dilutions are examined.

## References

BARNES E.M. (1956). Methods for the isolation of faecal streptococci (Lancefield Group D) from bacon factories. *J. Appl. Bact.* **19,** 193.

BARNES E.M. (1964). Distribution and properties of serological types of *Streptococcus faecuum, Streptococcus durans* and related strains. *J. Appl. Bact.* **27,** 461.

BLACKBURN T.H. & HOBSON P.N. (1962). Further studies on the isolation of proteolytic bacteria from the sheep rumen. *J. Gen. Microbiol.* **29,** 69.

BRYANT M.P. (1959). Bacterial species of the rumen. *Bact. Rev.* **23,** 125.

CLARKE R.T.J. & DI MENNA M.E. (1961). Yeasts from the bovine rumen. *J. Gen. Microbiol.* **25,** 113.

EL-SHAZLY K. & HUNGATE R.E. (1965). Fermentation capacity as a measure of net growth of rumen microorganisms. *App. Microbiol.* **13,** 62.

HOBSON P.N. & PURDOM M.J. (1961). Two types of xylan-fermenting bacteria from the sheep rumen. *J. Appl. Bact.* **24,** 188.

HOBSON P.N., MANN S.O. & SMITH W. (1962). Serological tests of a relationship between rumen selenomoads *in vitro* and *in vivo. J. Gen. Microbiol.* **29,** 265.

HOBSON P.N., MANN S.O. & SUMMERS R. (1966). Bacteria counting with a Coulter Counter. *J. Gen. Microbiol.* **45,** 5 p.

HUNGATE R.E. (1966). *The Rumen and its Microbes.* Academic Press, New York and London.

*Identification Methods for Microbiologists.* Part A. (1966). Academic Press, London and New York.

MADRCK T.F. & BARNES E.M. (1962). The distribution of Group D streptococci in cattle and sheep. *J. Appl. Bact.* **25,** 159.

MADRCK T.F. & BARNES E.M. (1958). A note on the growth of *Streptococcus bovis* in thallous acetate and tetrazolium media. *J. Appl. Bact.* **21,** 79.

DE MAN J.C., ROGOSA M. & SHARPE M.E. (1960). A medium for the cultivation of lactobacilli. *J. Appl. Bact.* **23,** 130.

MUSHIN R. & ASHBURNER F.M. (1964). Studies on coli-aerogenes and other Gram-negative intestinal bacteria in various animals and birds. *J. Appl. Bact.* **27,** 392.

PURSER D.B. & MOIR R.J. (1959). Ruminal flora studies in the sheep. IX. *Aust. J. Ag. Res.* **10,** 555.

ROGOSA M. (1964). The genus *Veillonella. I.J. Bact.* **87,** 162.

ROGOSA M., MITCHELL J.A. & WISEMAN R.F. (1951). A selective medium for the isolation and enumeration of oral lactobacilli. *J. Dent. Res.* **30,** 682.

SMITH H.W. (1965a). Observations on the flora of the alimentary tract and factors affecting its composition. *J. Path. Bact.* **89,** 95.

SMITH H.W. (1965b). The development of the flora of the alimentary tract in young animals. *J. Path. Bact.* **90,** 495.

VAN UDEN N., CO CARMO SOUSA L. & FARINHA W.M. (1958). The intestinal yeast flora of horses, sheep, goats and swine. *J. Gen. Microbiol.* **25,** 113.

# PART V. A SMALL MOBILE LABORATORY FOR RUMEN MICROBIOLOGICAL WORK

P N HOBSON, S O MANN and R SUMMERS

In the preceding chapters are described methods for study of rumen and intestinal microorganisms and their metabolic activities in field experiments. These necessitate a laboratory near the animals from which samples are taken, and we give a brief description of a mobile laboratory which we have built and used for all the work described in the preceding chapter. The laboratory could also be used for other microbiological fieldwork, and with the addition of a second van equipped for sterilizing media could form the basis of a complete, independent, field unit. A more detailed description of the van will be published elsewhere, or may be obtained from the authors.

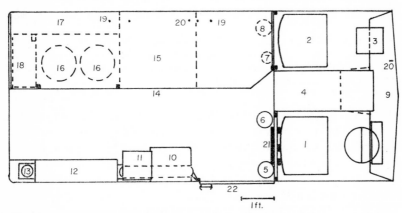

Figure 12.1. Sketch plan of mobile laboratory floor area, showing main furniture and apparatus.
1. Driver's seat.  2. Passenger seat.  3. Extension to dash-board for microscope, working position.  4. Folding table, working position; back of engine housing underneath.  5. $CO_2$ cylinder.  6. $N_2$ cylinder.  7. Vacuum pump.  8. Spare $CO_2$ cylinder.  9. Dash-board.  10. Sink.  11. Water tank for sink, spare wheel underneath.  12. Shelves above wheel arch.  13. Thermos flasks, on racks one above the other.  14. Edge of bench.  15. Drawer unit under bench.  16. Propane cylinders.  17. Wheel arch below bench, containing battery and power point, jack, wheel brace, etc.  18. Box below bench.  19. Propane and $CO_2$ gas taps.  20. Flexible stem lights.  21. Travelling position of folding table and 3 jacks clipped upright behind driver's seat.  22. Sliding side door, arrow showing direction of motion.  Scale indicator=1 ft.

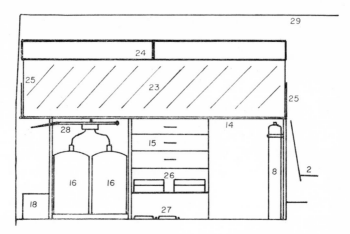

Figure 12.2. Sketch elevation of left-hand side of mobile laboratory, showing main features.
Numbers up to 22, as Fig. 12.1.    23. Peg board back to bench.    24. Two 6″ wide shelves above bench.    25. Perspex draught screens at ends of bench.    26. Culture tube tacks, spring-clamped to shelf.    27. Bases for jacks in slots on shelf.    28. Spade. 29. Roof line of van.    Scale as Fig. 12.1.

The van is a Renault Estafette, 15 cwt, high-roof van without structural alterations. The standard van gives 6 ft internal headroom and is fitted with side and end doors. However, the dynamo is replaced by a 12 volt alternator charging a 12 volt battery used to power the engine and also fitted with a take-off for an electricity supply to the bench. A fluorescent light is fitted on the roof over the bench.

An 8 ft long bench is fitted on a square section steel frame to the near-side at a height of 35 in (Figure 12.1). Bench width is 28 in. Above this at roof height are fitted two shelves 8 ft × 6 in (Figure 12.2). The side of the van above the bench and below the shelves is covered with peg-board, painted with fire-resistant paint, on which apparatus can be fixed in spring clips. At each end of the bench are vertical Perspex sheets to help prevent draughts blowing the gas flames about. Underneath in the center is fitted a unit with three drawers ($3\frac{3}{4}$ in × 25 in × 22 in) and two shelves. On one side of the drawer unit are two 28 lb propane cylinders and a box for 'oddments' of apparatus (Figure 12.3) and on the other side the pieces of a Butchart balance are fixed to the side of the unit and there is space for an insulated box for specimens, buckets, working clothing, etc. The propane cylinders are

Figure 12.3. Inside of van photographed through rear doors.
The apparatus is in the 'working position' with water tanks taken out.
The photograph was taken before the second shelf above the bench.
The spade and extra jacks were fitted.

connected via a reduction valve and permanent piping to gas taps on the bench. On the other side of the van the spare wheel is fitted near the door and above this is a small sink and a water container with tap. Above the rear wheel arch, which contains the petrol tank, is fitted a series of shelves carrying anaerobic jars, specimen jars, reagents and stains, a portable pH meter, vacuum flasks, etc.

Behind the driver's seat are spare jacks for stabilizing the van when it is in field use, a folding table which fixes between the front seats in the working position, a 22 lb cylinder of nitrogen and a 14 lb cylinder of pure (analytical grade) $CO_2$. This latter is coupled via a reduction gauge and permanent pipeline to taps on the bench, from where it can be used, with rubber tubing and bent hypodermic needles fitted with cotton-wool filters, for culture manipulation under anaerobic conditions. The nitrogen cylinder is fitted with a needle valve and tubing for filling anaerobic jars. Behind the passenger seat a hand-operated vacuum pump and a spare $CO_2$ cylinder are fitted. In front of the passenger seat and under the dash-board is a first-aid kit and a board which can be fitted, with bolts and wing nuts, on to the dash-board to make an extension on which a microscope can be bolted for use by a worker sitting in the passenger seat. A flexible stem map-reading light on the dash-board illuminates the microscope mirror, and there is also a light above the windscreen. A similar flexible light is fitted to the bench for microscopic work there. Two water baths heated by propane burners and controlled by a bimetallic strip thermostat through a 12 volt solenoid-operated gas valve are carried clamped on the bench and are used one at each end of the bench during experiments. Asbestos boards to protect the bench top and any bunsen burners in use go under these. A metal trough at one end of the bench serves as a container (filled with disinfectant if necessary) for used pipettes and glassware.

Two water containers, one of 40 litres capacity to fill the water baths, and one of 10 litres capacity of distilled water, are carried on the floor of the van strapped to the wheel arch and are taken out of the van during experiments.

**Apparatus carried, other than that mentioned above.** All apparatus is either fixed into racks, in compartmented drawers, fixed with spring toggles to shelves, or in clips above the bench, or clamped to the floor.

Post mortem equipment, knives, bone saw.

Stomach tubes, pump and ancillary equipment.

$1\frac{1}{2}$ gallon plastic buckets, 2.

Balances. Butchars to 250 g, spring balances to 5 kilos.

Hand centrifuge for 10 ml tubes and McCartney bottles.
100, 250, 500 ml plastic measuring cylinders.
Plastic beakers, 50 ml to 1000 ml.
Thermos flasks, 1 pint and 2 pint wide mouth.
Plastic filter funnels, large (8 in) or small.
Sterile pipettes, 1, 5, 10 ml, 72 in containers.
Sterile Pasteur pipettes, 48 in box.
Wide mouth pipettes for sample measurement, non-sterile pipettes for reagents. 1, 5, 10 ml, 24.
Glass conical flasks, 50 ml, 10. 100 ml, 1.
All glass syringes, needles, and rubber bungs for fermentation tests, 6.
Clamps (6) and support to fit above water bath for fermentation.
Vials of substrates weighed out for fermentation tests, 24.
Bunsen burners, 3.
Gas ring, 1.
Tripods and gauzes, 2.
Metal jugs for boiling water to melt agar media, 2.
Spring clamps for holding tubes of media in boiling water, 6.
Racks for tubes of media. 4 × 36 tubes. On shelves under drawer unit.
Rack for holding 12 bottles of diluting fluids, on bench.
Plastic trays for rolling culture tubes in cold water, 2.
Platinum wire loops.
Rubber bladders for transferring gases from cylinders to cultures.
1 ml syringes for drop dilution counts on agar plates, 2.
Instrument sterilizer for boiling syringes.
Anaerobic jars, 2.
Thermometers, $-10°$, $-110°$, 2.
Battery operated pH meter and buffers. Indicator papers.
Plastic bottles of reagents for fermentation tests, for fixing samples for microbiological and biochemical tests, 200 ml, 6.
Plastic dropping bottles of microscopic stains, 6.
Staining rack to go over sink.
Coplin jars, 3.
Plastic specimen jars, 250 ml, 27. 1 litre, 2.
Sterile McCartney and Universal bottles for samples, 40 each.
Sterile saline in plastic container.
Polythene bags for specimens, 24 assorted sizes.
Plastic specimen bottles, 25 ml narrow neck, 6.

Microscope, slides, cover slips, immersion oil, lens tissues. Watchmaker's lens.

Box for microscope slides of specimens.

Diamond pencil, felt-pen marker, pencils, pens, notebooks.

Sheelts filter paper and grease-proof paper, gauze-cloth roll.

Cotton wool, absorbent and non-absorbent.

Plastic self-adhesive labels, string, cellotape, fabric adhesive tape.

Rubber teats, artery and ordinary forceps, scissors, scalpels, spatulas, screw clips, rubber tubing.

Stop watch.

Rubber gloves soap, towels, dusters. Toilet rolls in holders under bench for tissue for wiping bottles, etc.

Fire extinguisher.

Water-proof torch and spare batteries.

Spade, wheel chains, spare fan belt, tools for van and for apparatus.

Folding seats for workers travelling in back of van, 2.

The disposition of the main apparatus, shelving etc., is shown in the figures. The van floor is covered with heavy rubber matting over felt.

## PART VI. UREA RECYCLING

### A T PHILLIPSON

Read (1925) observed that camels eating a low-nitrogen diet excreted very small quantities of urea in their urine. This was confirmed by Schmidt-Nielsen, Schmidt-Nielsen, Houpt and Jarnum (1957) who also observed the urinary clearance of camels on a low-nitrogen diet was depressed and that the same was true for the sheep (Schmidt-Nielsen and Osaki, 1958). They suggest that urea retention was due to its transfer to the rumen where it might serve as a useful source of nitrogen for the bacterial population therein. The transference of nitrogen to the rumen in the saliva secreted by a single parotid gland was found by Somers (1961a and 1961b) to be about 0·3 g daily in animals consuming about 15 g daily. Of this, 60—70% was accounted for by urea. When 3 g of urea (1·4 g of urea nitrogen) was infused into the blood while the animals were eating, an extra 0·12 g of urea nitrogen appeared in the parotid saliva secreted daily and about 1·0 g of urea nitrogen daily in the urine. The same experiment applied to sheep consuming 6·8 g of nitrogen daily and consequently in a negative nitrogen balance showed that about an extra 0·3 g of urea nitrogen appeared in the parotid saliva daily and only an extra 0·3 g in the urine. In both experiments the portion unaccounted for, other than that present in the other salivary secretions, presumably diffused into the alimentary tract. Such a diffusion has been found in the rumen of sheep (Houpt, 1959; Ash and Dobson, 1963), and the quantity diffusing seems to be substantially greater than the quantity entering in the parotid saliva. Presumably, urea will also diffuse into the other parts of the stomach and the intestine as well, as occurs in rabbits, rats, mice, and man. The slow infusion of urea into the rumen of cattle fed on straw has been shown to increase the digestibility of the straw by some 10%, and to improve appetite. In addition, the animals passed from a negative to a small positive nitrogen balance (Campling, Freer and Balch, 1962).

The retention of urea by ruminants maintained on low nitrogen rations is influenced by the state of hydration of the animal, for in sheep (Schmidt-Nielsen and Osaki, 1958) and in cattle (Livingston, Payne and Friend, 1962) the ingestion of water and diuresis is accompanied by increased renal clearance of urea. However, even in sheep which have access to water *ad libitum* (Somers, 1961*b*) a low nitrogen diet resulted in greater urea retention than occurred when a diet adequate in nitrogen was consumed.

Urease activity in the rumen is sufficient to ensure the rapid hydrolysis of urea entering that organ to ammonia and carbon dioxide (Pearson and Smith, 1943). Ammonia nitrogen is known to be incorporated into the protoplasm of bacteria in a form that is precipitable by trichloroacetic acid (Phillipson, Dobson, Blackburn and Wilson, 1962) while many rumen bacteria fail to grow *in vitro* unless ammonia nitrogen is present in the media (Bryant and Robinson, 1961).

The use of urea as a partial substitute for protein in cattle feeds, especially when the dietary supplies of nitrogen are low, is successful providing a suitable supplement of a starch-rich food or molasses is given as well to provide an easy source of fermentable carbohydrate. The addition of urea as such to all roughage diets is less successful, and this is probably because the structures of plant fibres are not fermented sufficiently rapidly. Deamination of urea occurs so rapidly that most of the ammonia so derived is absorbed rapidly, reconverted to urea in the liver and, except for any recirculated to the rumen, is excreted in the urine. The physiological supply of urea to the rumen is spread over the 24 hours in a fairly even way and under these circumstances the opportunity for a higher proportion of ammonia derived from it being assimilated by bacteria is greater than if a large quantity is ingested in a short time. The experiments quoted above show that slow infusions of urea to the rumen of animals fed on roughages do improve the nutritional state of the animal.

## Methods

**1.** The ability of an animal to conserve nitrogen can be tested roughly by making collections of their urine when they are in a negative nitrogen balance and examining the quantity of urea excreted during 24 hours. Urea should be estimated as such rather than total nitrogen which includes the endogenous products creatine and creatinine. If 24-hour collections are impossible, a comparison of urea to total nitrogen in samples of urine will provide some indication of the animal's ability to conserve urea. The state of hydration of the animal will probably influence the results, and both hydrated and de-hydrated animals in a negative nitrogen balance should be investigated.

**2.** Probably the best and simplest method of testing an animal's ability to recirculate urea to the alimentary tract is to give small, slow infusions of urea labelled with $^{14}C$. Urea entering the alimentary tract is deaminated and the quantity of $^{14}C$ excreted as carbon dioxide by the animal will serve as a measure of the extent to which this occurs. This experiment has been done by Leifer, Roth and Hempelmann (1948) and by Dintzis and Hastings (1953) on mice, by Walser and Bodenlos (1959) on man, and by von Engelhardt and Nickel (1965) on goats.

In order to test the maximum passage of urea to the alimentary tract, the experiment should be carried out on animals in a negative nitrogen balance and in both the dehydrated state and the normal hydrated state.

As this experiment needs laboratory facilities for collecting expired gases, urine (preferably) and also for counting $^{14}C$, it can only be contemplated when these are available.

**3.** Cannulation of the parotid duct allows the collection of saliva and its content of urea to be determined. A simple way of passing fine Polythene tubing into the parotid duct via the parotid papilla of the anaesthetized animal and the collection of saliva are described by Ash and Kay (1959). The tube is exteriorized by threading it on to a needle which is passed through the cheek anterior to the papilla. Collections of saliva for 24 hours or longer can be made. It can be accepted that the secretion from one gland over a period of 24 hours fairly closely represents the gland on the other side of the head except that during rumination the gland on the side of the mouth on which the bolus is chewed secretes more rapidly than the gland on the opposite side.

Repetition of Somers' (1961*b*) experiment to determine how much of urea given slowly as an infusion into the blood stream, preferably over a period of not less than 1 hour, appears in the parotid saliva will give some indication of the ability of the animal to recirculate urea to the rumen by this route. Again this experiment should be done while the animal is in a negative nitrogen balance and the effects of dehydration compared to the animal in a normally hydrated state.

Sodium depletion will occur if the saliva collected is not returned to the animal. This will result in (a) a decrease in the quantity of saliva secreted and (b) a decrease in its sodium content and an increase in its potassium content. To prevent this change the animal should be given sodium bicarbonate by mouth in quantities that are sufficient to replace sodium lost in the saliva. If sheep or cattle have access to salt or to drinking water containing a weak solution of sodium bicarbonate, they will consume voluntarily an amount sufficient to maintain a normal sodium balance with only a small variation.

### References

ASH R.W. & DOBSON A. (1963). The effects of absorption on the acidity of rumen contents. *J. Physiol.* **169**, 39–61.

ASH R.W. & KAY R.N.B. (1959). Stimulation and inhibition of reticulum contractions, rumination and parotid secretion from the forestomach of conscious sheep. *J. Physiol.* **149**, 43–57.

BRYANT M.P. & ROBINSON I.M. (1961). Studies on the nitrogen requirements of some ruminal cellulolytic bacteria. *Appl. Microbiol.* **9**, 96–103.

CAMPLING R.C., FREER M. & BALCH C.C. (1962). Factors affecting the voluntary intake of food by cows. 3. The effect of urea on the voluntary intake of oat straw. *Br. J. Nutr.* **16**, 115–124.

DINTZIS R.Z. & HASTINGS A.B. (1953). The effect of antibiotics on urea breakdown in mice. *Proc. Nat. Acad. Sci. Wash.* **39**, 571–578.

ENGLEHARDT W.V. & NICKEL W. (1965). Die permeabilität der pansenwand für harnstoff, antipyrin und wasser. *Pflugers Arch ges. Physiol.* **286**, 57–75.

HOUPT T.R. (1959). Utilization of blood urea in ruminants. *Am. J. Physiol.* **197**, 115–120.

LEIFER E., ROTH L.J. & HEMPLMANN L.H. (1948). Metabolism of $C^{14}$-labeled urea. *Sci.* **108**, 748.

LIVINGTON H.G., PAYNE W.J.A. & FRIEND M.T. (1962). Urea excretion in ruminants. *Nature, London* **194**, 1057–1058.

PEARSON R.M. & SMITH J.A.B. (1943). The utilization of urea in the bovine rumen. 2. The conversion of urea to ammonia. *Biochem. J.* **37**, 148–153.

PHILLIPSON A.T., DOBSON M.J., BLACKBURN T.H. & WILSON M. (1962). The assimilation of ammonia nitrogen by bacteria of the rumen of sheep. *Br. J. Nutr.* **16**, 151–166.

READ B.E. (1925). Chemical constituents of camel's urine. *J. Biol. Chem.* **64**, 615–617.

SCHMIDT-NIELSEN B., SCHMIDT-NIELSEN K., HOUPT T.R. & JARNUM S.A. (1957). Urea excretion in the camel. *Am. J. Physiol.* **188**, 477–484.

SCHMIDT-NIELSEN B. & OSAKI H. (1958). Renal response to changes in nitrogen metabolism in sheep. *Am. J. Physiol.* **193**, 657–661.

SOMERS M. (1961a). Factors influencing the secretion of nitrogen in sheep saliva. 1. The distribution of nitrogen in the mixed and parotid saliva of sheep. *Aust. J. Exp. Biol.* **39**, 111–128.

SOMERS M. (1961b). Factors influencing the secretion of nitrogen in sheep saliva. 4. The influence of infected ures on the quantitative recovery of urea in the parotid saliva and the urinary excretions of sheep. *Aust. J. Exp. Biol.* **39**, 145–156.

WALSER M. & BODENLOS L.J. (1959). Urea metabolism in man. *J. Clin. Invest.* **38**, 1617–1626.

# PART VII. WATER METABOLISM

## R N B KAY

The need for water in regulating body temperature is discussed in the chapter on environmental physiology. This section will briefly summarize other aspects of water intake and loss and will indicate some measurements that may usefully be made. Many aspects of the water economy of animals in the desert have been fully considered by Schmidt-Nielsen and Schmidt-Nielsen (1952), Schmidt-Nielsen (1964), and Macfarlane (1964). The water economy of mammals in general has recently been reviewed by Chew (1965) and an earlier review relating to farm animals is that by Leitch and Thompson (1944).

### Water intake

This is the sum of the water drunk, the water contained in the food, and the water released by oxidation (metabolic water). The amount of water drunk by domestic ruminants that are not under heat stress depends largely upon how much food is eaten and on its dryness, and on the drinking occurring at mealtimes. Conversely, restriction of water may lead to a reduction in food consumption (Clark and Quin, 1949; Balch, Balch, Johnson and Turner, 1953, French, 1956). Under temperate conditions non-lactating cattle and sheep need some 2 or 3 l of water per kg of dry matter eaten. Animals on a succulent pasture may not drink any water at all and the same is true of wild animals in East Africa during the wet season. During the dry season animals that need to drink, such as buffalo, wildebeest, zebra and elephant, congregate where permanent water is available, and the area in which they can forage is determined by how frequently they must return to drink. Impala, Grant's gazelle, gerenuk and oryx, on the other hand, can live in waterless areas, feeding mainly at times of day when the vegetation is moistened by dew and absorbed water (Lamprey, 1963). Metabolic water, amounting in weight to roughly half the intake of digestible dry matter, can make an important contribution to water resources and the kangaroo rat can survive on it indefinitely (Schmidt-Nielsen, 1964).

An animal's water intake determines its water turnover rate. The measurement of this is described in the next section (Determination of body fluids in the field). The water turnover rate varies greatly from species to species. Cattle, for example, have a rapid turnover, camels a slow one, while sheep and kangaroos are intermediate (Macfarlane, Morris and Howard, 1963). Breeds of cattle differ in their water economy; zebus are able to conserve water and maintain appetite better in the face of water deprivation than are European breeds (Phillips, 1961).

### Water stores in the gut

At one time it was supposed that the camel was able to store water in the 'water cells' of its stomach. It has now been shown that these sacculations serve no such function (Schmidt-Nielsen 1964). Nevertheless, the rumen fluid as a whole acts as a reservoir when the animal is deprived of water or food. Sheep reduce the volume of the rumen fluid by 2 or 3 l during the first two days of deprivation and so delay other effects of dehydration (Macfarlane, Morris, Howard, McDonald and Budtz-Olsen, 1961; Hecker, Budtz-Olsen and Ostwald, 1964).

### Effects of dehydration

These differ greatly between animals (Schmidt-Nielsen, 1964). Man sweats water (and salt) freely and reduces the urinary excretion of water and salt. The panting sheep loses only water and so reduces the urinary excretion of water but increases that of salt. Panting also reduces the $CO_2$ tension of the blood and so causes alkalosis. The osmotic pressure of the blood rises considerably in some species, digestive secretions diminish and the faeces become dryer, appetite fails, and the volume of the blood is reduced leading to an increase in the viscosity of the blood. These last changes occur relatively early during dehydration in man, and the resulting impairment of the circulation to the skin prevents the dissipation of heat produced in the deep tissues so that a sudden and fatal rise in body temperature occurs. The camel, on the other hand, maintains its blood volume even when it has lost a quarter of its weight through dehydration and so withstands dehydration much better (Macfarlane *et al.*, 1963).

**Drinking behaviour**

A thirsty Australian aborigine is said to drink enough water at a single draught to rehydrate himself fully. This is perhaps an acquired skill and certainly a rare one; Europeans and most other mammals generally consume rather less than their full water deficit at their first drink and make good the remainder in the next few hours. This suggests two of the mechanisms by which water intake is regulated; a coarse metering of intake that allows thirst to be satisfied temporarily before more than a little of the water consumed can have been absorbed, and a fine control which operates only when the water has been fully absorbed. In the first category buccal and pharyngeal receptors have been shown to be important in sheep (Bott, Denton and Weller, 1965); the second level of control is related to the osmotic pressure of the body fluids.

It is a great advantage, of course, for animals to be able to drink their fill rapidly and leave exposed and dangerous places such as water holes quickly. But the sudden consumption of a large volume of water leads to many changes (Macfarlane, unpublished): salts first pass into the digestive tract leading to a fall in blood volume and antidiuresis and then follows rapid absorption of water, a fall in blood osmotic pressure, and compensatory movements of salts and water between the other body fluids. In sheep, water may be absorbed so rapidly as to bring about some haemolysis.

If the drinking water is saline, a larger volume must be drunk than of pure water to allow for the greater urine volume needed to excrete the additional salt load. Magnesium or sulphate may cause diarrhoea, so adding to water losses. Species vary in their tolerance (Pierce, 1957); saline water is clearly of no use if its osmotic pressure is greater than the maximum attainable in the urine, and in practice must be considerably less than this. Sheep regulate their water intake relative to salt intake so as to maintain a urinary concentration of about 1 osmolal (Macfarlane, unpublished). The kangaroo rat produces such a concentrated urine that it can survive on sea water (Schmidt-Nielsen, 1964). Birds and reptiles that excrete concentrated saline solutions from their nasal glands have a special advantage.

## Water losses

Water is lost by evaporation (*see* Section 4, Chapter 4) and in the faeces and urine. The concentration of water in the faeces varies characteristically between species and breeds. It decreases in dehydrated animals and in animals consuming little indigestible matter. Generally, however, faecal water excretion will be proportional to faecal dry matter excretion and some herbivores such as cattle may lose about as much water in their faeces as in their urine.

Obligatory losses of water in the urine depend on the amount of solute to be excreted and the maximal concentrating ability of the kidney tubules. The solutes concerned are principally potassium and sodium salts from the food and water, and end products of nitrogen metabolism such as urea. Large amounts of potassium may be lost in the faeces, but little sodium can be eliminated by this route. Urea is unimportant when the food contains little nitrogen since much of the small amount of urea that is produced is re-absorbed from the kidney tubules and passes back to the digestive tract. This process is described in the previous section (Urea recycling).

Secretion of milk entails an additional water loss in lactating animals and growing or pregnant animals also necessarily have a slightly greater water requirement to permit a positive water balance.

## Watering regimes for captive animals

Usually it will be enough to provide an excess of fresh water once daily or as a continuous supply. The water can serve as a useful vehicle for administering urea or salts, as is practised in certain feeding systems. If the diet of ruminant animals is so poor as to endanger their nitrogen balance, there may be some advantage in restricting the water intake, since this may reduce their urinary nitrogen losses (*see* Urea recycling).

## Methods of study

Many aspects of the water economy of wild or tame animals may be studied in the field. Simple measurements often yield interesting results, especially of a qualitative or comparative nature. The interpretation of these measurements can then be aided and extended by laboratory studies under carefully defined conditions.

**Field studies.** The rate of water turnover may be studied by the methods described in the next section. This indicates total water intake and is related to species differences, to the heat load, to season and the water content of the diet, to availability of drinking water and the frequency of drinking, to reproduction and lactation, etc. The amounts drunk by tame animals may be found either by direct measurement of the water supplied or by weighing the animal. Knowledge of the total amount of water in the body and its distribution between intracellular and extracellular compartments, especially plasma, under different environmental conditions tells one a great deal about the state of dehydration of the animal and how it contends with dehydration.

*Killed animals.* Much can be learned from freshly killed animals. Blood may be sampled from the heart and examined for haematocrit, red cell fragility, osmotic pressure, specific gravity, etc. Heparin should be used to prevent coagulation, rather than citrate or oxalate. Urine and faeces can be obtained from bladder and rectum. Urine osmotic pressure and faeces water content indicate the state of hydration of the animal, and the route of excretion of various salts and nitrogenous compounds may be determined. The volume and composition of the contents of various compartments of the gut indicate the size of the gut fluid space. Botanic analysis of food residues indicates the nature of at least the less digestible fraction of the diet. In the case of small animals, total body water can be measured directly by drying.

*Tame animals.* Blood samples can be drawn for analysis from tame animals in the field under light restraint. It is advisable to catheterize the jugular vein or some other superficial vein an hour or two in advance since venesection causes the spleen to contract and this rapidly affects the blood; even in very tame animals capture and handling may have similar consequences (Turner and Hodgetts, 1959). Urine can be obtained by catheterization in the female or by more devious means, and faeces by 'grab-sampling' from the rectum.

Some of these measurements, such as haematocrit, osmotic pressure and specific gravity of plasma and fragility of the red cells, weights or volumes of gut contents, etc., can be made immediately with equipment that can be carried in a lorry. Analysis for salts and water content, which will not change with storage of the sample, can be made without difficulty in the laboratory.

**Intensive studies.** There is not space to consider the more intensive experimental study of water metabolism. Many such studies have been made; for example, in man (Adolph, 1947), camels (Schmidt-Nielsen, Schmidt-Nielsen, Houpt and Jarnum, 1956), sheep (Macfarlane *et al.*, 1961), and cattle (Bianca, Findlay and McLean, 1965). Laboratory studies, in particular, have the advantage that they enable changes caused by dehydration or rehydration, say, to be studied in great detail and under carefully controlled conditions or at least under defined conditions, for example, of climatic stress and solute load. They therefore allow one to compare various physiological aspects of water metabolism in different species under conditions which are not complicated by uncertain differences in environment, water intake, diet or behaviour. New measurements are also possible. Water balance can be measured directly, for example; and by using calorimeter chambers heat losses can be assessed and pulmonary and cutaneous water losses can be measured separately.

### References

ADOLPH E.F. (1947). *Physiology of Man in the Desert*. Interscience, New York.
BALCH C.C., BALCH D.A., JOHNSON V.W. & TURNER J. (1953). Factors affecting utilization of food by dairy cows. 1. The effect of limited water intake on the digestibility and rate of passage of hay. *Br. J. Nutr.* **7**, 212.
BIANCE W., FINDLAY J.D. & MCLEAN J.A. (1965). Responses of steers to water restriction. *Res. Vet. Sci.* **6**, 38.
BOTT E., DENTON D.A. & WELLER S. Water drinking in sheep with esophageal fistulae. *J. Physiol.* **176**, 323.
CHEW R.M. (1965). Water metabolism of mammals. *Physiological Mammalogy*, Vol. 2, edited by MAYER W.V. & VAN GELDER R.G. Academic Press, New York and London.
CLARK R. & QUIN J.I. (1949). Studies on the water requirements of farm animals in South Africa. 1. The effect of intermittent watering on Merino sheep. *Onderstepoort J. Vet. Sci.* **22**, 335.
FRENCH M.H. (1956). The effect of infrequent water intake on the consumption and digestibility of hay by Zebu cattle. *Emp. J. Exp. Ag.* **24**, 128.

HECKER J.F., BUDTZ-OLSEN O.E. & OSTWALD M. (1964). The rumen as a water store in sheep. *Aust. J. Ag. Res.* **15**, 961.

LAMPREY H.F. (1963). Tarangire Game Reserve, Tanganyika. *E. Afr. Wildl. J.* **1**, 63.

LEITCH I. & THOMPSON J.S. (1944). The water economy of farm animals. *Nutr. Abst. Rev.* **14**, 197.

MACFARLANE W.V. (1964). Terrestrial animals in dry heat: Ungulates. *Handbook of Physiology*, Sec. 4, Dill D.B. Am. Physiol. Soc. Wash.

MACFARLANE W.V., MORRIS R.J.H. & HOWARD B. (1963). Turn-over and distribution of water in desert camels, sheep, cattle and kangaroos. *Nature, Lond.* **197**, 270.

MACFARLANE W.V., MORRIS R.J.H., HOWARD B., McDONALD J. & BUDTZ-OLSEN O.E. (1961). Water and electrolyte changes in tropical Merino sheep exposed to dehydration in summer. *Aust. J. Ag. Res.* **12**, 889.

PHILLIPS G.D. (1961). Physiological comparisons of European and Zebu steers. 2. Effects of restricted water intake. *Res. Vet. Sci.* **2**, 209.

PIERCE A.W. (1957). Studies on salt tolerance of sheep. 1. The tolerance of sheep for sodium chloride in the drinking water. *Aust. J. Ag. Res.* **8**, 711.

SCHMIDT-NIELSEN B., SCHMIDT-NIELSEN K., HOUPT T.R. & JARNUM S.A. (1956). Water balance of the camel. *Am. J. Physiol.* **185**, 185.

SCHMIDT-NIELSEN K. & SCHMIDT-NIELSEN B. (1952). Water metabolism of desert mammals. *Physiol. Rev.* **32**, 135.

SCHMIDT-NIELSEN K. (1964). *Desert Animals. Physiological Problems of Heat and Water.* Clarendon Press, Oxford.

TURNER A.W. & HODGETTS V.E. (1959). The dynamic red cell storage function of the spleen in sheep. 1. Relationship to fluctuations of jugular haematocrit. *Aust. J Exp. Biol. Med. Sci.* **37**, 399.

# PART VIII. DETERMINATION OF BODY FLUIDS IN THE FIELD

## W V MACFARLANE

In the previous section, aspects of the water economy of large herbivores was discussed. In this section, methods of measuring quantity and turnover of body water will be described.

## Identification

Mark the animal for ready identification and recapture. This may be done with branding fluid, such as the lanolin-based Siromark, or with luminescent tape.

## Weight

A spring balance hung from a tree may be used. Smaller animals such as sheep and goats can be weighed in a sack hung from the balance. Larger animals require to be hoisted by a block and tackle system unless there is a mobile weighbridge. A canvas sheet passed under the belly allows hoisting, or tying of the legs and hoisting upside down may be used. The inverted hoist by the legs is not safe for animals over 200 kg.

## Intravenous injection

The jugular is the most readily available vein in ruminants and a 12-gauge needle through which plastic tubing can be passed provides a satisfactory route. For most of the larger animals it is desirable to infiltrate the region with 1 ml of 2% Xylocaine (lignocaine), since in camels and cattle, for instance, it is easier to penetrate the hide when a long tapered scalpel has been used to cut the fibrous tissue. The cannula is sutured in place firmly and a nylon obturator is passed through the blood vessel. Braun manufactures plastic cannulae with hollow steel needles within them (B Braun, Carl Braun Str. 1, Melsungen, W. Germany). These allow cannulation with one penetration of the vein, and then the needle is withdrawn. If the cannula is to be left in place, the flange requires removal, since it tends to catch on standing objects and will be dragged out.

## Fluid spaces

**Rumen volume.** Theoretically, any marker that mixes with rumen contents should yield an estimation. In practice it is not easy to obtain good mixing of a representative sample. Polyethylene glycol is useful, a dose of 0·1 to 0·2 g/kg body weight being given by mouth and rumen samples taken at 2, 4, and 6 h (Sperber, Hyden and Ekman, 1953). Downes and McDonald (1964) found similar answers with $^{51}$Cr-EDTA. Phenol red gives a similar answer (Hecker, Budtz-Olsen & Ostwald, 1964), in which phenol red in water (1 ml/kg) is given by mouth and samples taken hourly for 7 h.

**Plasma volume.** Injections of Evans Blue dye ($T_1824$) intravenously, about 0·3 mg/kg body weight, provide an albumen marker for plasma volume.

Some samples of dye have a purple contaminant. Usually the Evans Blue made by Gurr is satisfactory (G. T. Gurr, 136–144 New Kings Road, London, S.W.6, England). Although the most reliable results are obtained by taking samples at 20, 40, and 60 min. after injection and extrapolating to zero time, a sample taken 5 min. after injection gives the same answer with 3—5% reliability. Estimations of $T_1824$ in the field require a centrifuge in order to provide suitable plasma although serum can serve as long as haemolysis does not intervene.

**Extracellular fluid volume.** This is satisfactorily estimated by thiocyanate although the bromine space gives a similar answer. Thiocyanate spaces are somewhat larger than those of inulin and they approximate the sodium space. Tropical ruminants normally have a considerably larger thiocyanate space than those of the temperate zone.

The injection of 10—15 mg NaSCN/kg body weight intravenously provides a suitable marker substance and a blood sample after 20 min. in sheep, cattle and camels gives an estimate which is as reliable as extrapolation to zero time (Macfarlane, Morris, Howard & Budtz-Olsen, 1959). It is preferable to separate plasma in the field for this estimation although serum can be used.

**Total body water and water turnover.** These are most readily estimated by the use of tritiated water (Morris, Howard & Macfarlane, 1962). A dose of 3—5 uCi/kg body weight given intramuscularly or intravenously provides a suitable marker if a sensitive tritium-counting instrument is used. Equilibration of the tritiated water with all body fluids including those of the rumen requires about 2 h in lambs, 6 h in sheep, 7 h in cattle, 10 h in camels and longer in animals which have been dehydrated, since the salivary and pancreatic flow of fluids is reduced in this condition. The fall in concentration in samples of blood or urine taken at 3—5 day intervals after equilibration gives the turnover rate, and in hot climates the turnover is 3—6 times faster than in cool areas.

Deuterium oxide may also be used. A mass spectrograph is desirable if reasonable accuracy of determination of concentration is to be obtained and it is best to administer at least 0·5 ml/kg.

**Precautions.** It is best to prevent the animal from consuming any water for 2—3 h before injection, since a large intake of water is quickly excreted and gives a falsely high estimate of total body water. No fluid or food should be allowed until the equilibration sample has been taken, since some fluid could be not completely equilibrated with TOH.


### Estimations


**Phenol red** is centrifuged from rumen contents at 25,000 rpm and estimated in NaOH: $Na_2HPO_4$ buffer pH 12 at 5560 Å. The absorption at pH 3 is used as a blank and subtracted from that at pH 12 (Hecker *et al.*, 1964).

**Evans Blue** is estimated photometrically in plasma by comparison with a known dilution of dye in plasma. It is read at a wavelength of 6200 Å. An emulsion of fat will reduce transmission, and haemolysis makes the estimation difficult. Extraction of the blue dye from plasma by alcohol in these circumstances gives a more satisfactory reading.

**Thiocyanate** is estimated by the method of Bowler (1944) after precipitation of plasma protein with trichloracetic acid. Colour is developed with ferric nitrate and estimated photometrically relative to a standard at a wavelength of 4600 Å. Serum or heparinized plasma should be used and oxalate avoided.

**Tritium.** The specific activity of samples is readily estimated by the methods of Vaughan and Boling (1961). Sublimation *in vacuo* using liquid nitrogen separates tritiated water from electrolyte and chromogens which would interfere with scintillation counting. A counting medium made up of PPO, POPOP, naphthalene and dioxane will accept water up to the ratio of about 1:5 scintillation fluid.

The total body water is derived from the known amount of tritium and the concentration at equilibrium. The water turnover derives from the half time of tritium in the body which is related to the proportion of the total body water turned over daily, by the factor 0·693:

$$\frac{0\cdot693 \times 100}{t_{\frac{1}{2}}} =$$

percentage of water pool turned over daily where $t_{\frac{1}{2}}$ = biological half life of TOH.

If the volume of fluid in the body changes between the first and last tritium sample, a correction is needed. A further injection of tritium or consumption of a known volume of water will provide the second volume estimation. Correction formula for change of volume of body water:

$$- bt = 2{\cdot}03 \log \frac{V_t}{V_o}$$

where $b$ is rate of change of volume, $t$ is time interval, $V_o$ is initial volume, $V_t$ is final volume. This allows correction of the apparent turnover rate.

Regression lines relating body solids and fat content to the body water content of sheep have been provided by Panaretto (1963). Such correlations must be made for each species. Among ruminants the volume of gut contents is not easily determined so that fat content is less easily estimated, although deprivation of food and water for 24—48 h brings the gut content to a more uniform fraction of the whole.

### Milk intake and water turnover

During a period in which young ruminants are drinking milk and not eating other foodstuffs, water turnover of the young is a direct function of milk intake. Turnover comprises water from milk and metabolic water derived from milk.

In lambs, injection of 5 uCi of TOH/kg body weight allows estimation of total water content and therefore of body solids as well as of water turnover. Since the young use about 250 ml/kg/day, the biological half life of TOH is short (about 3 days). The animals are also growing rapidly so that a further measurement of body water is required at the end of 10 to 14 days in order to determine the expansion of the total bodywater. Correction for expansion of this space in relation to turnover is indicated under Estimations.

### References

BOWLER R.B. (1944). The determination of thiocyanate in blood serum. *Biochem. J.* **38**, 385.
DOWNES A.M. & MCDONALD I.W. (1964). The chromium-51 complex of ethylenedramine tetraacetic acid as a soluble rumen marker. *Br. J. Nutr.* **18**, 153.
HECKER J.F., BUDTZ-OLSEN O.E. & OSTWALD M. (1964). The rumen as a water store in sheep. *Aust. J. Ag. Res.* **15**, 961.
MACFARLANE W.V., MORRIS R.J.H., HOWARD B. & BUDTZ-OLSEN O.E. (1959). Extra cellular fluid distribution in tropical Merino sheep. *Aust. J. Ag. Res.* **10**, 269.

MORRIS R.J.H., HOWARD B. & MACFARLANE W.V. (1962). Interaction of nutrition and air temperature with water metabolism of Merino wethers shorn in winter. *Aust. J. Ag. Res.* **13**, 320.

PANARETTO B.A. (1963). Body composition *in vivo*. III. The composition of living ruminants and its relation to the tritiated water spaces. *Aust. J. Ag. Res.* **14**, 944.

SPERBER L, HYDEN D.S. & EKMAN J. (1953). The use of polyethylene glycol as a reference substance in the study of ruminant digestion. *LantbrHogsk. Ann.* **20**, 337.

VAUGHAN B.E. & BOLING E.A. (1961). Rapid assay procedures for tritium-labeled water in body fluids. *J. Lab. Clin. Med.* **57**, 159.

# PART IX. YIELD AND COMPOSITION OF MILK

## J H TOPPS

If possible, the length of lactation should be defined and the daily yield and composition of the milk measured at frequent intervals during this period. A *minimum* of 6 to 8 samples per lactation is probably necessary to obtain some assessment of lactational yield and to provide a sufficient number of samples for analysis to show any variation in chemical composition of the milk.

### Measurement of yield

It is usually very difficult or virtually impossible to hand-milk undomesticated animals. The dam often refuses to 'let down' her milk, so little or no milk can be withdrawn. To overcome this difficulty the animal may be injected with oxytocin to facilitate milking or a more indirect method of measuring milk yield may be used. For example, the offspring can be weighed before and after suckling. If this is done on several occasions during the day at times when suckling would occur under natural conditions, a reasonably accurate determination of yield may be acheived. Unfortunately, the use of this method does not provide samples for chemical analysis. (*See* Hartman, Ludwick and Wilson, 1962, for a description of this technique with sows.) Braude and Mitchell (1950) have shown that both the amount of milk released and the length of time of flow of milk are related to the concentration of oxytocin in the blood. The amount of hormone to be used, therefore, will vary with species and possibly with body weight within a species. The injection of the optimal amount should allow the complete withdrawal of milk by hand. However, the amount of milk withdrawn may exceed that obtained by the 'sucking and weighing' technique.

Recently it has been found that certain wild animals can be hand-milked when they are immobilized by M99 or Orpavine. For details of this method see the monograph by Harthoorn (1965).

## Preservation of samples

If possible, analysis of the milk should be carried out on the day it is obtained. When analysis is delayed, duplicate samples (each approx. 250 ml) should be placed in sealed containers, a preservative added, and, if possible, surrounded by a foamed plastic container with solid carbon dioxide. Choice of a preservative will be governed by its effect on milk constituents. To minimize these effects the use of an antibiotic may be necessary. Other preservatives include potassium dichromate and formalin (1 ml per litre of milk), which are, however, unsuited for electrophoretic or protein studies.

## Measurement of composition

Standard methods of milk analysis should be adopted (*see* Ling, 1948). The determination of constituents which are present in particularly high levels may not be possible by some standard procedures. For example, with milk very rich in butter-fat the Gerber method cannot be used for the determination of this component. In these cases, either other methods must be used or, at times, new methods of analysis may have to be developed. For methods to determine the vitamin content of milk, *see* Gregory, Ford and Kon (1958) and Thompson, Heny and Kon (1964).

### References

BRAUDE R. & MITCHELL K.G. (1950). Let-down of milk in the sow. *Nature, Lond.* **165,** 937.
GREGORY M.E., FORD J.E. & KON. S.K. (1958). The B-vitamin content of milk in relation to breed of cow and stage of lactation. *J. Dairy Res.* **25,** 447.
HARTHOORN A.M. (1965). *Wildl. Monog.* 14. Wildl. Soc. Wash. D.C.
HARTMAN D.A., LUDWICK T.M. & WILSON R.F. (1962). Certain aspects of lactation performance in sows. *J. Anim. Sci.* **21,** 883.
LING E.R. (1948). *A Textbook of Dairy Chemistry*, Vol. 2. Chapman & Hall, London.
THOMPSON S.Y., HENRY K.M. & KON S.K. (1964). Factors affecting the concentration of vitamins in milk. 1. Effect of breed, season and geographical location on fat-soluble vitamins. *J. Dairy Res.* **31,** 1.

*Chapter 12*

# PART X. TOXIC PLANTS AND TOLERANCE TO PLANT TOXICITIES

## N W PIRIE

Toxic plants are so widespread on unimproved grazing and browsing land that many textbooks and instruction manuals deal with their distribution, compositions, and physiological effects. In the context of this handbook there is no need for any further survey of that theme. The narrower theme of the different susceptibilities of animal species, or of individuals within a species, to poisoning by specific plants is, however, relevant; this is needed to call attention to the fact that land which would be hazardous to one species of herbivore can provide good fooder for another, and to the possibility that useful tolerant varieties could be selected from with a species of animal in which most individuals risk poisoning by the local fodder. This aspect of toxicity seems not to have been systematically surveyed; I am not an expert in it and will welcome information about further examples and comments on the examples I cite.

It is a matter of common observation that goats eat weeds rejected by other animals, but it is not clear how much they eat nor whether more discriminating herbivores would be poisoned if they could be induced to eat the quantity eaten by goats. Similarly, the apparent tolerance of sheep for ragwort (*Senecio*) arises because they reject it (Forsyth, 1954; King, 1961); and the apparent susceptibility of horses to yellow star thistle (*Centaurea solstitialis*) could depend on the greater unwillingness of other species to eat it even when fodder is sparse (Cordy, 1964; Mettler and Stern, 1963). Careful quantitative work is therefore needed before confident statements can be made, and the mere survival of one species in an environment in which another dies may mean little. Furthermore, a plant that is being rejected is likely to outgrow those that are being eaten and this not only wastes space but puts stock permanently at risk should their selectivity change. Thus horses and cattle usually reject buttercups, bryony, rhododendron and some other species but may suddenly become addicted, and keep to the addiction even after a bout of poisoning.

Acorns are toxic to most species but their value as pig-feed is both traditional and well established experimentally. Kudu eat *Synadenium capulare*, *Strophanthus* sp. and *Euphorbia* sp. toxic to other species. The last is particularly interesting because the juice of the edible *Euphorbia* blinds kudus when rubbed on the eye. Hyrax and chicken (Sale, 1965) eat *Phytolacoa dodecandra* which is toxic to most animal species. Fluoracetic acid, the toxic component of *Dichapetalum cymosum*, has been thoroughly studied because of its potentialities as a weapon and rodenticide; the mean lethal dose ranges from 0·2 mg/kg for dogs, coyotes and sheep to 10 or more for monkeys and fowls (Foss, 1948; Chenoweth, 1949; Meldrum, Bignell and Rowley, 1957).

The mode of action of fluoracetic acid is understood but not the reason for these species differences. Tentative explanations for other differences can sometimes be given. For example, the rumen organisms of deer, unlike those of cattle, are not inhibited by the oil from white cedar leaves (Short, 1963). Bracken contains an enzyme that destroys thiamine (vitamin B1) and so is particularly toxic to animals that lack the vitamin-synthesising flora of the rumen. But ruminants succumb to an injury of the bonemarrow (Evans, Humphreys, Goulden, Thomas and Evans, 1963); and non-ruminants, protected from vitamin deficiency by B1 injection, get tumors (Evans and Mason, 1965). Detailed analysis has therefore complicated rather than simplified the position. With photosensitivity, or hypericism, the position is relatively clear. Whether it arises from feeding buckwheat (*Fagopyrum esculentem*) or from weeds such as bog asphodel (*Narthecium ossifragum*) the effects depend on sunlight; they are minimised in dark skinned or heavily fleeced animals and prevented by keeping the animals in the dark. The external growth on a fodder of organisms such as *Sporidesmium* can damage the liver so that there is an aberration in bile pigment or chlorophyll metabolism leading to photosensitivity. Similarly, the coumesterol content of a fodder seems to be greatly increased by insect damage or plant virus infection. An apparently harmful forage may not, therefore, be harmful *per se*.

Intraspecific variations in sensitivity are well known in pharmacology and commonly invoked t  explain the death of only part of a herd, but there is generally no evidence that all the animals ate the same amount of material (e.g. the study of poisoning by trematol from some *Compositae*—Christensen, 1965). All rabbits are relatively resistant to *Atropa belladonna* and some can eat the plant in bulk. This ability depends on variation in the amount of the enzyme atropinase (Sawin and Glick, 1943); the ability to produce the enzyme is inherited as in incomplete dominant and a homozygous strain of rabbits was bred. It would be interesting to have similar studies in other species and with other toxic materials.

Finally there is adaptation or mithridatism. Many species can be slowly adapted to the ricin in castor bean press-cake (Clarke, 1947). Mice and rats, but not dogs and rabbits (Chenoweth, 1949), become tolerant to fluoracetic acid; and it is claimed that sheep become tolerant to the mimosine in *Leucena glauca*. These phenomena would repay fuller study but the effect is by no means general, for many other poisons are cumulative.

### References

BIOCHEM. SOC. COLLOQUIM. Toxic substances present in plants and their effect on grazing animals. (1963). *Biochem. J.* **88**, 55.

FORSYTH A.A. (1954). British poisonous plants. *Minn. Ag. Fish Fd. Bull.* 161.

GARNER R.J. (1961). *Veterinary Toxicology*. Bailliere, Tindal & Cox, London.

KING J.O.L. (1961). *Veterinary Dietetics*. Bailliere, Tindal & Cox, London.

WILLIAMS R.J. (1963). *Biochemical Individuality* (reprint 1956). Wiley, New York.

CHENOWETH B. (1949). Manofluoracetic acid and related compounds. *Pharmocol. Rev.* **1**, 383.

CHRISTENSEN W.I. (1965). Milk sickness. A review of the literature. *Econ. Bot.* **17**, 293.

CLARKE E.G.C. (1947). Poisoning by castor seed. *Vet. J.* **103**, 273.

CORDY D.R. (1954). Nigropollidal encephalomalacia in horses associated with ingestion of yellow star thistle. *J. Neuropath.* **13**, 330.

EVANS I.A., HUMPHREYS D.J., GOULDEN L., THOMAS A.J. & EVANS W.C. (1963). Effects of bracken rhizomes on the pig. *J. Comp. Path. Ther.* **73**, 3.

FOSS G.L. (1948). The toxicology and pharmacology of methyl flouracetate (MFA) in animals with some notes on experimental therapy. *Bri. J. Pharmacol.* **3**, 118.

MELDRUM G.K., BIGNELL J.T. & ROWLEY I. (1957). The use of sodium flouracetate (compound 1080) for the control of the rabbit in Tasmania. *Aust. Vet. J.* **33**, 186.

METTLER F.A. & STERN G.M. (1963). Observation on the toxic effects of yellow star thistle. *J. Neuropath.* **2**, 164.

SALE J.G. (1965). Hyrax feeding on a poisonous plant. *E. Afr. Wildl. J.* **3**, 127.

SAWIN P.B. & GLICK D. (1943). Atropinesterase, a genetically determined enzyme in the rabbit. *Proc. Nat. Acad. Si., Wash.* **29**, 55 p.

SHORT H.L. (1963). Rumen fermentations and energy relationships in white-tailed deer. *J. Wildl. Mgmt.* **27**, 184.

# 13

# Habitat Manipulation

## PART I. REGIONAL SURVEYS FOR HABITAT STUDIES

### C S Christian

Terrestrial herbivores can adjust their body metabolism to some degree and thereby adapt themseves to a variety of conditions. Nevertheless, their natural distribution and productivity are restricted or favored by features of the landscape having significance to food and water supplies, shelter, presence of predators, and natural hazards and other factors. In most cases, however, the 'habitat' of a species usually represents a variety of habitats when expressed in terms of landscape characteristics rather than a single landscape type. Within the broad geographic limitations imposed on species by accessibility or climatic extremes, there will be a range of situations which a species may select throughout the day, season, or year.

The terrain of an area is composed of a number of land forms which recur in association with one another in such a way as to produce a distinctive landscape pattern referred to as a land system (Christian and Stewart, 1953; Christian, 1958). Each land system is the product of a long history of landscape formation. On the one hand, there has been the operation of a particular sequence of geomorphic processes acting at certain rates and for particular periods of time over an area characterized by a distinctive geological material; on the other, there has been an accompanying biological development varying with the climate and the landscape development. Of practical significance is the fact that these land systems, which incidentally can readily be recognized in most cases on aerial photographs, do exist and can soon be recognized by the seeking eye in the field. Moreover, their boundaries are usually marked by changes in geological boundaries, or by boundaries of major geomorphological processes such as wind or water erosion, peneplenation, uplift, faulting, alluviation, partial stripping, flooding, or a combination of these. Each land system represents a distinctive habitat; or, more usually, a distinctive complex of habitats significant to the distribution and productivity of herbivorous species.

The distinctive land forms which compose each land system may themselves be subdivided further into land units, similar land units being repeated at each occurrence of the land foim within the land system. Each land unit is characterized by certain observable features: shape or slope, a particular array of soils or rocks constituting the surface mantle, vegetation communities and associated fauna. Allied to these will be a characteristic array of microclimates within the broader climatic type in which the land system occurs and characteristic surface and sub-surface hydrological regimes. Collectively, all these features of the land unit constitute the total environment at that site. It is this whole complex which determines whether or not the land unit is an appropriate natural habitat for a given species.

While it will be the ultimate objective of the ecologist to analyze environment and to understand the impact of individual factors on the survival, growth, and reproduction of a species, this may more often be done successfully as a second step after an initial understanding of distribution and productivity in relation to total environments has been gained.

Terrestrial herbivores may roam many land units or land systems. A study of the distribution and behaviour of species in relation to those areas which they occupy continuously or for certain periods of their life cycle is an essential first step towards identifying the major control factors within the environment influencing survival, growth, reproduction, and behaviour.

The situation with managed species, be it commercially or otherwise, is a special case. In such instances, the environment can be subject to deliberate modification. Unproductive environments may in many cases be made productive, or productive environments destroyed. The region survey describing the inherent features of the landscape, provides a basis for selecting either those areas which are naturally well suited to the introduced species, or those which are potentially the best and easiest to improve. Alternatively, if a form of management or land treatment is already known by experience or experiment to have a certain effect, desirable or deleterious, on a given land unit or series of units, the regional survey will indicate the similar areas where the treatment can be expected to give a like response. It thus provides a basis for estimating potential for land use development, but in new areas, decisions may often have to wait the conclusion of a research or experimental program initiated for this purpose.

Land resource surveys, which aim at recognizing, describing, mapping and assessing these natural units of the landscape, have been developed in a number of countries, but particularly so in Australia. They are being used for a variety of purposes ranging from studies with an engineering objective (Grant, 1965) to forestry studies (Sprout, *et al.*, 1966) and tropical land development (Mabbutt *et al.*, 1965). Their application to mammal ecology has not been specifically developed but they provide a more systematic approach to procedures which form an essential part of an ecologist's study. A general review of the concepts and methodology, and an extensive bibliography, was presented to the UNESCO Conference on Principles and Methods of Integrating Aerial Survey Studies of Natural Resources for Potential Development at Toulouse in September, 1964, and this should soon be available in print (Christian and Stewart in press). A brief description of the method was also given by Christian (1959).

Such resource surveys are concerned with a number of features of landscape, each of which represents a specialist field of study. Hence, the ideal survey approach is to use a team of specialists, the essential members of which would be a geologists and/or geomorphologist, a pedologist, and a plant ecologist. Such a team would need also to have access to climatological data and analyses and, preferably, the help or advice of an ecoclimatologist. Subsequent assessments of the environment may require further specialist help, particularly in the fields of hydrology, general agriculture, and animal husbandry, if the intention is to manage resources or to improve them for animal production purposes.

In Australia, teams of this nature have been operating within the Division of Land Research of CSIRO* for many years. Not all countries have organizations of this kind established for extensive regional surveys, but it would be well within the scope of universities, by collaboration between a number of university departments to establish similar teams of people interested in the same region or project. In other circumstances, it may be possible to develop them through collaboration between a number of government departments.

The basic team works together in the field and there is considerable advantage gained from the integrated study of the landscape in this way as compared with a series of independent studies.

Not all animal ecologists, however, will be able to draw on the resources of teams as described. There is no really satisfactory alternative, as one research worker can hardly be competent in so many fields. This is especially relevant in areas not previously studied. Where one person has to operate alone, his best approach is to make prior studies of such maps and information as are available dealing with the individual components of the environment, and to

* Commonwealth Scientific & Industrial Research Organization.

consult with specialists in order to arrive at the most significant subdivisions of the area under study in respect to each feature. By comparing these various subdivisions with field observations, and with aerial photographs if they are available, he will soon develop a sense of recognition of significant units of landscape. It will be necessary, however, for him to be familiar with at least the basic principles of description and classification of rocks and soils, and to learn how to take respresentative samples which can be referred to specialists in these subjects for interpretation. As vegetation is such an important feature of mammal habits, it will be essential for the animal ecologist to be able to recognize the main plant species of the area and to develop contacts for the determination of characteristic specimens of the less prominent species, which may well include poisonous or other wise deleterious species of considerable significance. He will need to develop a system of description of vegetation which is meaningful both in terms of habitat characteristics and in differentiating between plant communities, which vary with other features of the environment even though some of the dominant species and general structure may appear to be the same (Christian and Perry, 1953).

The animal ecologist must always bear in mind that the environment is complex and that there are many interactions between and interdependence of factors presenting cumulative or complementary consequences. Apparent relationships between animals and specific factors of the environment may, in fact, be mere reflections of the other factors. For example, the distribution of rabbits in relation to certain soils in arid Australia is thought to be more a matter of heat exchange properties in the rabbit burrows in the soil profile than any other feature. Such a factor may not retain the same significance in a different climatic zone.

The ecologist must be particularly careful in assessing the potential value of an area for increased animal production. The possibilities of changing the habitat by management of the animals and the vegetation, by changing soil fertility, and by the introduction of different plant species is so great that the potential production may bear little relationship to production in the natural state (Riceman, 1948; Specht, Rayson and Jackman, 1958; Davies and Eyles, 1965). In this respect, special care must be observed in drawing conclusions based on homoclimatic studies; climatic comparisons can be wildly misleading unless the total plant-soil-climate relationship is examined as a whole. The warning is equally true, of course, in the interpretation of any one of this trio of dominant environmental factors.

### References

CHRISTIAN C.S. (1958). The concept of land units and land systems. *Proc. Ninth Pacific Sci. Cong.* 1957. **20**, 74–81.

CHRISTIAN C.S. (1959). The eco-complex in its importance for agricultural assessment. In *Biogeography and Ecology in Australia.* Ser. Monog. Biol. 8 p. 587–605.

CHRISTIAN C.S. & PERRY R.A. (1953). The systematic description of plant communities by the use of symbol. *J. Ecol.* **41**, 100–105.

CHRISTIAN C.S. & STEWART G.A. (1953). General report on survey of Katherine-Darwin region, 1946. *C.S.I.R.O. Aust. Land Res. Ser.* 1.

CHRISTIAN C.S. & STEWART G.A. (in press). (1964). Methodology of integrated surveys. *Proc. UNESCO Conf. Principles and Methods of Integrating Aerial Survey Studies of Nat. Resources for Potential Development, Toulouse, Sept.,* 1964.

DAVIES J.G. & EYLES A.G. (1965). Expansion of Australian pastoral production. *J. Aust. Inst. Ag. Sci.* **31**, 77–93.

GRANT K. (1965). Terrain features of the Mt. Isa—Dajarra region and an assessment of their significance in relation to potential engineering land use. *C.S.I.R.O. Aust. Soil Mechanics Section. Tech. Paper* 1.

MABBUTT J.A. *et al.* (1965). Lands of the Port Moresby Kairuku area, Territory of Papua and New Guinea. *C.S.I.R.O. Aust. Land Res. Ser.* 14.

RICEMAN D.S. (1948). Effect of zinc, copper and phosphate on subterranean clover and lucerne grown on Laffer sand, near Keith. *Bull Commonw. Sci. Ind. Res. Org.* 234.

SPECHT R.L., RAYSON P. & JACKMAN M.E. (1958). Dark Island heath (Ninety-mile Plain, South Australia). VI. Pyric succession: Changes in composition, coverage, dry weight, and mineral nutrient status. *Aust. J. Bot.* **6**, 59–88.

SPROUT P.N., LACATE D.S. & ARLIDGE J.W.C. (1966). Forest land classification survey and interpretation for management of a portion of the Niskonlith Provincial Forest, Kamloops District, B.C. *Columbia For. Ser. Tech. Publ. T*60. Dept. of For. Publ. 1159.

# PART II. SOIL AND BIOLOGICAL PRODUCTIVITY

## V A KOVDA and I V YAKUSHEVSKAYA

*Editorially re-written by* R G CLEMENTS

### Introduction

A close correlation has been established between the properties of soils and the productivity of plants. Thus it is possible to equate the productivity of soils with the quantity of plant material produced per unit time per unit area. The humus reserves of the soils (transformed plant material and energy involved) and the quantity of microorganisms and animals which inhabit the soil may serve also as an indirect indicator of the biological productivity of soils.

The productivity of animals, particularly herbivores, depends directly on the productivity of plants. Interactions between plants and animals are more complicated than those between soils and plants due to the variations introduced by particularities of animal life, such as migrations, and by the activity of man, for example, the destruction of plants, animals and natural landscapes. Since the productivity of animals is indirectly related to soils through plant productivity, it is possible to speak of the parallelism of soil productivity and zoobiomasses of the natural landscapes only in a general way.

The biological productivity of some natural landscapes of the USSR may serve to illustrate these general relationships (Table 13.1). Soils of the southern taiga are acid derno-podzolic soils of low fertility as compared to the more fertile soils of the broad-leaf forests and the highly fertile chernosem soils of the meadow steppes. Total biological productivity can be increased considerably when cultivated. Likewise, under the influence of man the animal biomass can be increased greatly.

Any analysis of the role of soils in the biological productivity of plants and animals must consider highly complex properties of the soil, such as the different levels and balance of nutrients. The variability in biological productivity of soils is influenced by the geographical position (the quantity of solar energy and the moisture regime); the altitude (redistribution of food elements, moisture and heat in the correlated soil-geochemical landscapes); and, finally, the stage of soil development which, in turn, is influenced by the landscape evolution (topography).

## Division of soils

The general classification* of soils of the continents by thermal and moisture regimes is straightforward and generally accepted. Less well known is the classification of soils by ' pedo-geochemical formations't. This classification is based upon grouping soils having similar chemical and mineralogical properties under certain climatic regimes. The major formations based upon this classification are given in Table 13.2.

Within each formation the classification scheme recognizes the general forms of soil combinations found within geochemically connected landscapes‡. These are: bioaccumulative (eluvial), transit, hydrobioaccumulative,

---

* For a general review on the subject of soil classification, the reader is referred to a review presented in *Soil Science* (1949). Also various classification schemes are compared by Papadakis, 1964.

† This is similar to the classification set forth by Thorp and Smith (1949).

‡ See Bushnell (1945).

TABLE 13.1. Annual biological productivity of some natural landscapes of USSR
(Dry matter)

| | LANDSCAPE | | | | | |
|---|---|---|---|---|---|---|
| | South Taiga | | Broad leaf forest | | Meadow steppe | |
| | Biomass[1] | Energy reserves[2] | Biomass[1] | Energy reserves | Biomass[1] | Energy reserves |
| Higher plants | 8·5 | $15 \times 10^{10}$ [2] | 11·0 | $18 \times 10^{10}$ [2] | 11·0 | $18 \times 10^{10}$ [2] |
| Microorganisms and algae | 0·7 | $0·6 \times 10^{10}$ [3] | 1·9 | $1·5 \times 10^{10}$ [3] | 1·9 | $1·5 \times 10^{10}$ [3] |
| Vertebrates | 0·0017 | | 0·0025 | | 0·0030 | |
| Invertebrates | 0·86 | | 0·90 | | 1·44 | |
| Humus reserves in one meter layer of soil | 99 | $37 \times 10^{10}$ [3] | 215 | $66 \times 10^{10}$ [3] | 549 | $168 \times 10^{10}$ [3] |

[1] Biomass is in tons per hectare.
[2] Energy reserves are calories per hectare from data of Rabinovitch E.I. 1951.
[3] Energy reserves are calories per hectare from data of Aliev S.A. 1966.

TABLE 13.2. Pedo-geochemical formations.

| | Formation | Great group | Climate | Typical minerals |
|---|---|---|---|---|
| I | Acid allitic soils | bocsitic allitic | humid rainy tropics | boemite, gibbsite, hydragillite |
| II | Acid allitic-kaolinitic soils | kaolisol ferruginous | humid tropics | kaolinite, gibbisite goethite |
| III | Acid kaolinitic soils | krasnozems red earths zheltozems-yellow earth rubrozems | humid tropics | hydromicas, kaolinite, vermiculite, residual primary minerals. |
| IV | Acid siallitic soils | podzols podzolic brown podzolic brown forest lessive | humid subboreal and boreal belts | hydromicas, kaolinite, vermiculite, residual primary minerals |
| V | Neutral and slightly alkaline siallitic soils | cinnamon, grey forest mediterranean | moderately arid sub-tropics temperate climate | poligorskite, chlorite, montmorillinite, calcite |
| VI | Alkaline montmorillinite humid soils | chernozems meadow brunizems vertisols grumosol | different climate, but mostly semiarid or periodically arid climate of savannas, steppes and prairies | montmorillinite, calcite (aragonite) sometimes gypsum |
| VII | Alkali and saline soils | solonetzs solonchaks sodic soils | different climate, but mostly arid or periodically arid | montmorillinite, calcite, gypsum, semihydrate, galite, mirabillite, tenardite |
| VIII | Volcanic soils on ashes | andosol hydrol soil curobocu trumao | different climates | primary minerals, especially volcanic glasses, palagonite, allophanes |

and hydroaccumulative. Descriptions of each of the soil combinations are presented in the following paragraphs.

**I. Eluvial** landscapes with automorphic (bioaccumulative) soils are typical of elevated relief in watersheds with deep laying ground waters which do not influence the contemporary soil-forming processes. The geochemical removal of matter with simultaneous biogenic accumulation in the humus layers predominate. Since the total balance of elements is negative, we expect a deficit of some plant nutrients.

Very often in these soils we can recognize the features of former hydro-morphism, which must be taken into consideration because they change considerably the total supply of nutrients and energy. Examples of these features are:

(a) considerable amounts of montmorillonitic or amorphous clay and compactness of the soil mass.

(b) presence of secondary formations of silica powdering, opal-crusts, layers, flows, segregations, concretions and pseudomorphism

(c) formation of hardpans, cemented with clay or with $R_2O_3$, $SiO_2$ or $CaCO_3$

(d) presence of grey mottled horizons in the middle part or on the bottom of the soil profile indicating poor internal drainage

(e) accumulation of manganese, iron films, concretions, and ortsteins in the soil profile

(f) accumulation of carbonate in the soil

(g) presence of gypsum and easily-soluble salts

(h) remnants of hydromorphic flora and shells of mollusks.

These signs are mostly peculiar to the soils of ancient alluvial plains and dry deltas, which have been raised up by the neotectonic processes.

**II. Transit** landscapes of the terraces and slopes which are characterized by migrations of soil particles and their subsequent sorting according to their degree of mobility. This type of landscape may be divided into two subtypes:

(a) transit-eluvial with the predominance of removal processes. The ground-waters do not take part in the soil formation of this type. The accumulation of elements which are brought by the side-flow is insignificant and the soil differs very little from those of the eluvial landscape.

(b) transit-accumulative landscapes which are characterized by the influence of ground-water on the lower part of the soil profile and on the root systems of plants. This creates an additional supply of nutrients and moisture for plants and creates conditions for the biogenic accumulation in the upper part of the soil profile. There occurs simultaneously a sedimentation of a considerable amount of elements from the deluvial runoff. As a whole, the balance of elements has an accumulative character, in spite of the leaching process which takes place in the upper part of the soil profile. Such landscape is characterized by semi-hydromorphic soils with a high productivity.

**III. Hydrobioaccumulative** landscapes (underwater) which are characterized by accumulation of soil material and chemical substances from the water suspensions and solutions and accumulation of organic matter as a result of activity of numerous living organisms inhabiting the water basins. These landscapes may be considered as a sort of peculiar geochemical barrier for the minerals and organic matter which are brought by the hydrographic network. Here occurs a primitive amphibious soil forming process—the presoil stage. The formation of these primitive underwater soils goes on simultaneously with the formation of sedimentary parent rocks.

Thus, the soils within each pedo-geochemical formation are characterized by a definite type of exchange (input-output) and nutrient levels. Types of exchange are shown in Figure 13.1. This general form of soil combinations can be applied to soils of different thermal belts (climatic regimes). Thus, for example, the pedo-geochemical formation of alkaline montmorillinitic humus soils are found on the savannas, prairies, pampas and northern steppes. While the chernozems dominate the watersheds and the meadow-chernozems, dark-coloured soils and vertisols are found on the lowlands, river terraces and flood plains.

Quite naturally, the nutrient level and moisture regimes in soils of the related elementary pedo-geochemical landscapes influence their biological productivity. For example, the annual synthesis of dry organic matter by higher plants on the watershed areas of the southern taiga was 8·5 tons per hectare as compared to 32 tons per ha on the flood terraces in the zone; the biomass of microorganisms and algae was 0·7 tons per ha versus 4·2 tons per ha; and the zoobiomas was 0·36 versus 0·76 tons per ha.

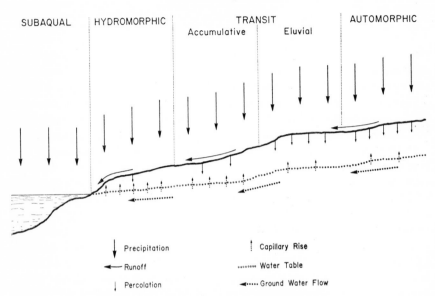

Figure 13.1. Elementary pedo-geological landscapes.

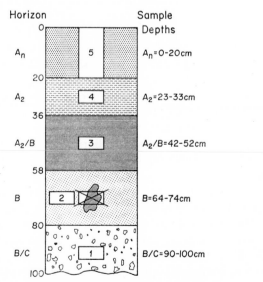

Figure 13.2. Sampling of soils by genetic horizons.

## Soil survey

One of the first considerations in the initiation of soils investigations is the need for the survey and classification of the soil of the area. Usually general surveys of the area are available, but it may be necessary to give a more detailed survey depending upon the scope and objectives of a given investigation. Classification of the soils by pedo-geochemical formations enables one to establish the general characteristics of the soils, i.e., bioaccumulative, hydrobioaccumulative, transit or hydroaccumulative. The nature of the classification yields general information about the soils of each category; such as, probable pH range, exchange capacity, per cent base saturation, drainage, etc. Then this may be followed by a more detailed study of the soil within each formation, which will yield information on profile characteristics and the chemical, physical and mineralogical properties of the soils.

In detailed investigations, several soil geomorphological cross sections which include the varieties of biogenocenoses of elementary pedogeochemical landscapes must be done. The soil profiles must be deep enough (up to 3–4 meters) to permit the observation and sampling of all genetical horizons of soil, roots and parent rock. If possible, it is desirable to deepen the profiles with a hand borer for investigation of the lower layers of parent and bedding rocks.

On the lower terraces and lower elements of relief the profiles must reach the soil-ground-water level. The samples from wooded areas must include the forest litter, humus, and transitional horizons and rock. In a case where there is a bleached layer (the eluvial horizon—$A_2$) and a darker layer (illuvial B horizon) under the humus layer, samples must be taken from the humus layer (separately from the upper part, the most interwoven with the roots of plants) further on to the transitional layers and rock. In the swampy soils the samples must be taken from the gley horizons and from the peat layer, if present. In all soils samples of new formations, inclusions and ground waters must be taken separately.

Samples are taken from the middle part of each genetic horizon if the horizon has a thickness of at least 10 cm (*see* Figure 13.2). Sample size should be at least 500 g. Where special analyses are foreseen, the sample size should be increased accordingly. If the thickness of the horizon is great enough, several samples should be taken from each one.

For obtaining the reliable analytical data, at the same time with the taking of samples from the main soil profiles, the mixed samples (for arable) and individual ones (for the virgin soils) are taken, but only from the upper humus and ploughed layer. The amount, depending on the homogeneity of soils and kind of analysis, varies from 3—7 to 20 samples on an area of approximately 3–5 hectares.

The test procedure of taking the samples is as follows: The samples are formed from 5 soil samples taken on a small area, 100 to 400 square meters. The first sample is taken from the main soil profile. The other four samples are taken in a cross-shaped way, 10—20 meters from the first sample (Figure 13.3). All samples (with a weight of 300—400 g each) are put on a sheet of

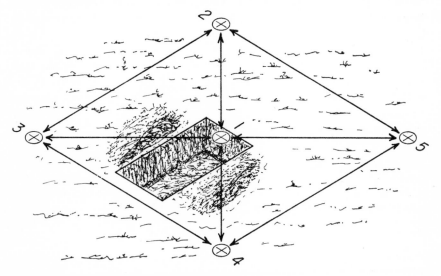

Figure 13.3. Soil sampling for a composite sample.

thick paper and mixed very thoroughly. Finally, an average sample of 300–400 g is taken.

On unploughed fields the individual samples are taken with a hand auger at the full depth of the humus horizon (if it is not greater than 20 cm.). When the thickness of the horizon is greater than 20 cm, two individual samples are taken from the upper and lower parts of the horizon.

The location, quantity, horizon, and depth of all samples must be recorded in a field journal, along with descriptions of each profile,

In addition to the description of the soil profile, it is necessary to give particular attention to the possible relic-marks connected with the former stages of evolution, for example, traces of hydromorphism. These indications are of great importance as they may influence the biological productivity.

The detailed investigation of a territory, besides including classification and detailed study of the soils, must also define other characteristics of the elementary landscapes, such as the character of the biogeocenoses. It is possible that one elementary eluvial landscape may consist of several different forest biogeocenoses—a cultural biogeocenose, a meadow biogeocenose, etc. Within a given biogeocenose, vegetation of different ages may vary considerably by several important indices. For example, the *Carex-Aegopodium dubraya* (oak forest) of the forest-steppe varies considerably depending upon the age of the stand. Two stands, 22 and 220 years old, are characterized by the following indices (*see* Table 13.3).

TABLE 13.3. Comparison of two stands of oak forest of different ages.

| | Age of stand | | |
|---|---|---|---|
| Index | 22 years | 220 years | |
| Organic matter reserves | 48·8 | 413·3 | tons/Ha dry matter |
| Litter reserves | 13·8 | 14·2 | tons/Ha dry matter |
| Quantity of nesting birds | 4·0 | 20·0 | pairs per Hectare |
| Earthworms | 80·0 | 35·0 | individuals/Sq M |
| Microorganisms | 6875·0 | 8335·0 | individuals × $10^3$/g of soil |
| Removal of elements by precipitation | 108·5 | 147·2 | Kg/Ha |
| Removal of elements by ground water | 77·1 k | 97·0 | Kg/Ha |
| Current increase of wood | 5·0 | 73·0 | tons per Ha |

## Chemical studies

The number of chemical studies depends on the scope of the investigation. However, certain studies are considered quite indispensable for describing the main characteristics of the soil. These are pH of the soil; water and salt solutions; redox potential (Eh); the level of the active forms of N, P, and K; absorbed cations; and easily soluble salts and humus.

The fluctuations of the average biomass and chemical compounds may be considerable—which is necessary to remember when studying the soil productivity of separate regions. This may be illustrated by some concrete examples.

The content of several macro- and micro-elements in soils of watershed areas and flood terraces of the same region of the southern taiga is shown in Table 13.4. When studying the microelements in soils formed on the moraine deposits of some regions in the northern part of the USSR, the soils of the watershed areas revealed a very sharp deficit of such important microelements as Cu, Co and Ni. At the same time the semihydromorphic and hydromorphic soils were relatively rich in the above mentioned microelements, and the levels were higher than those of the chernozem soils. The reserves of these elements in the sapropel of lakes (the subaqual landscapes) are so great that they are used as a source of micronutrient fertilizers.

The data on the total and mobile forms of microelements in soils are also of great interest, since several cattle diseases are connected with the deficiency or abundance of some of the microelements. Particular attention to the microelements must be given in those regions where the deficiency or abundance of microelements in soils is expected. Deficiency in Cu, Co, and I is often associated with sandy soils and areas of high rainfall and excessive leaching.

Local diseases of cattle and plants in landscapes deficient in some microelements are known in some parts of England, Latvia, and in the Kostroma and Yaroslavl regions of USSR. Examples of these are: an acobaltosis of animals which occurs when the cobalt content of the soil drops below a critical level (approximately 2–2·5 ppm of mobile Co); stag headness of fruit trees, and a 'disease of cultivation' of herbs, licking disease and anemia of cattle which are caused by the deficiency of Cu (lower than 1—3 ppm mobile Cu); and, finally, the widespread goiter endemia connected with the low content of Iodine (less than $10^{-5}\%$).

Abundance of microelements is very often connected with mountainous regions or with plain-lands with beds of fundamental rocks near the surface; that is, where the ore deposits are accompanied by the 'ring of accumulation' with an extremely high content of microelements.

Similar anomalys accompanied by the endemical diseases are described in Bashkir Republic (abundance of Cu) in the Aktjubinsk region (abundance of Ni and Cu), in Armenia (abundance of Mo) and in other districts.

The accumulation of more mobile microelements, for instance, of B, is also possible in continental lowlands under condition of an arid climate. A typical example of such a biogeochemical province is the Aralo-Kaspian lowlands, characterized by the accumulation of B and by associated endemical disease of cattle.

TABLE 13.4. Content of some macro and micro elements in the humus horizons of watershed and flood plain soils (total form).

| | % C | % N | % CaO | % Fe₂O₃ | % P₂O₅ | % SO₃ | ppm Mn | ppm Cu | ppm Co | ppm V |
|---|---|---|---|---|---|---|---|---|---|---|
| Derno-podzolic soil under forest | 0·55 | 0·08 | 0·25 | 0·26 | 0·03 | 0·14 | 1300 | 56 | 5 | 17 |
| Meadow soil of the central flood plain | 4·69 | 0·55 | 1·07 | 12·6 | 0·84 | 0·25 | 2200 | 70 | 34 | 140 |

In summary, the great diversity of soils, plants and animals in the natural landscapes necessitates interdisciplinary investigations which include studies of soil forming materials, geomorphology, underground and surface waters, soil water solutions, and higher and lower forms of plants and animals.

### References

*Agrochemical Cartography of Soils*. (1962). Acad. Sci. USSR. Moscow.
ALIEV A.A. (1966). Energy reserves, boundes in humus and biomass of microorganisms in soils of Azerbaijan and their bioenetical indexes. Theses of rept. to III All-Union delegate's meeting of soil scientists in Tartu.
BUSHNELL T.M. (1945). Some aspects of the catena concept. *Soil Sci. Soc. Amer. Proc.* **10**, 335–340.
DOBROVOLSKY G.B. (1964). Soils of river bottomland of upper and middle Volga basin. Doctor's dissertation. Moscow State Univ.
DOBROVOLSKY G.V. & YAKUSHEVSKAYA I.V. (1961). Several microelements content of bottomland soils of forest zone. Role of microelements in agriculture. *Trans. 2nd Inter-univ. Meet. Concerning Microelements*. Moscow State Univ.
GRETCHIN I.P., KAURITCHEV I.S., NIKOLSKY N.N. *et al.* (1964). *Practice on Soil Science*. Kolos. Moscow.
KOVDA V.A. (1966). Problems of biological and economical productivity of Earth. *Ag. Biol.* **1** (2) Moscow.
KOVDA V.A., FRIDLAND V.M., GLAZOVSKAJA M.A., LOBOZA E.V., ROZANOV B.G., ROZOV N.N., RUDNEVA E.N. & VOLOBUEV V.R. (1966). An assay of the legend construction to the 1: 5,000,000 World soil map. *Rept. Moscow Meet. Adv. Panel UNESCO/FAO World Map Proj.* Ed. 'MGU', Moscow.
KOVDA V.A. & YAKUSHEVSKAYA I.V. (In press). An assay of the Earth biomass estimation.
POLINOV B.B. (1956). *Teaching about Landscapes*. Selt. Trans. Acad. Sci. USSR.
PAPADAKIS J. (1964). *Soils of the World*. Buenos Aires. 141 p.
PRUCINKEVITCH Z. & PLICHTA V. (1965). Scientific problems of fertility of forest soils and criteria for its quantitative estimation. *Roczniki Gleboznawcae* **15** (2). Warszawa.
RABINOVITCH E.I. (1951). *Fotosynthesis*. Ed. I.L. Moscow.
SOIL CLASSIFICATION. (1949). *Soil Sci.* **67**, 77–230.
SUKATCHEV V.I. (1964). *Fundamentals of Forest Biogeocenology*. Nauka. Moscow.
THORPE J. & SMITH G.D. (1949). Higher categories of soil classification: Order, Suborder, and great soil groups. *Soil Sci.* **67**, 117–126.
VERNADSKY B.I. (1965). *Chemical Structure of Earth Biosphere and of its Surrounding*. Nauka. Moscow.
YAKUSHEVSKAYA I.V. (1965). Microelements in soils of moraine landscapes. Theses of repts. IV. All-Union Inter-Univ. Meet. on the prob. microelements and natural radioactivity of soils. Book 1. Petrozavodsk.

## PART III. WATER MANAGEMENT

### ROBERT E WILLIAMS

The ancient hunter waiting in concealment near the waterhold recognized the dependence of grazing animals on water. The livestock producer of the Great Plains and other steppe grasslands has always been acutely and sometimes painfully aware of the importance of water for animals.

Of all water sources, the hunter and stockman are most conscious of rain. Rain furnishes the life-giving moisture necessary to produce a forage crop. If rain fails, feed disappears. Livestock and game animals must be driven or migrate to greener pastures. If rain fails, streams, ponds, and lakes recede—the lesser drying up. Animals must find drinking water elsewhere or perish.

Location and spacing of natural and man-made waterings are a major influence in managing vegetation. Animal concentrations around stable water sources in arid or semi-arid country can denude the land. Erosion results and sediment is carried into the water, jeopardizing its quantity and quality.

Making effective use of available moisture for feed production and providing adequate, properly spaced waterings are two major objectives in managing rangelands and related grazing land for large herbivores.

One way to increase effectiveness of water for forage production is to make sure the water is absorbed by the soil. Water intake by soil is influenced by soil depth, soil texture and porosity, and the condition of the surface (Renner and Love, 1955). By leaving more herbaceous vegetation on the ground on overused ranges, moisture which enters the soil may be doubled or tripled. The extra water saved builds up the range more rapidly and increases the foreage supply.

Reducing those species of woody plants that produce little forage for big game or livestock makes more moisture available for desirable plants. In Texas it is estimated that non-useful woody plants use 40% of the state's total available moisture (Smith and Rechenthin, 1964). Research in Arizona showed that mesquite trees use 1725 pounds of water to grow one pound of dry matter and that catclaw acacia uses 2400 pounds. Desirable grasses such as sideoats grama use only 705 pounds of water per pound of dry matter; Arizona cottontop, 547; and blue grama, only 596 pounds.

Adequate, well-spaced water is important in managing grazing lands. Such water satisfies the needs of grazing animals and contributes to uniform and efficient use of forage.

Water requirements of animals vary greatly. Cattle should be watered every day for best results; horses and sheep thrive under somewhat less frequent waterings. In addition to kinds of stock, nature of forage and weather conditions determine livestock needs. In excessively steep or rough country, waterings for cattle should be about 1 mile apart. In flat country, water sources may be up to 5 miles apart (Stoddart and Smith, 1955).

Wild animals generally are more adapted to adverse water conditions, but do better when adequate water is available. The more limited the watering places, the greater the assemblage of animals near water. Rhinoceroses, waterbucks, and reedbucks stay close to water; the elephant travels 30 km. from water (Allee and Schmidt, 1951).

On the Wichita Mountain Wildlife Refuge in Oklahoma, the carrying capacity of the range for bison has been raised by building lakes and ponds (Fuller, 1960). Koford (1960) reported that vicuna range could be increased in Peru by constructing rock and earth dams in dry areas.

Kinds of constructed watering developments that may be feasible in a given area depend on: the permanent and intermittent natural water available; the productivity of the grazing area, climate, soil, and topography; and the materials, equipment, and labor available.

Careful initial planning, considering all available resources, information, and alternatives, is essential. Costly mistakes can be avoided through such planning (Hamilton and Jepson, 1940).

Constructed ponds and dugouts are widely used and, if conditions are favorable, offer a practical solution to many water problems. Topography, soil, runoff, evaporation, and animal requirements determine if a structure is feasible and whether it will furnish permanent or intermittent water.

Wells are expensive but are often justified because of their dependability and quality of water. Where springs exist, their development offers an efficient and economical opportunity for water development.

By using asphalt, plastics, or other impervious materials, land surfaces in small watersheds can now be treated to yield runoff water. Constructing storage tanks and watering troughs below such treated areas then makes water available for large grazing animals in areas where it was formerly impossible to do so because of soil or other conditions.

Troughs and tanks are expensive, but they are important in storing and using water in areas where water is scarce.

In addition to serving the needs of grazing animals, water developments serve many kinds of birds and small mammals. Managing man-made lakes and ponds for fish production is increasing.

Dual use of grazing lands by livestock and big game animals is not un-common. Income from hunting big game is greater than from livestock on some ranches.

Adequate, well-spaced water facilities contribute to sound management of ranches, public lands, wildlife refuges and game ranges for all large grazing and browsing animals.

#### References

ALLEN W.C. & SCHMIDT K.P. (1951). *Ecological Animal Geography.* 2nd Ed. John Wiley & Sons, Inc., New York. 715 p.

FULLER N.A. (1960). The ecology and management of the American bison. *Ecology and Management of Wild Grazing Animals in Temperate Zones.* Symposium, I.U.C.N. Warsaw, Poland.

HAMILTON C.L. & JEPSON H.G. (1940). Stock-water developments: wells, springs and ponds. *Farmers Bull.* 1859. U.S. Dept. of Ag.

KOFORD B. (1960). The ecology and management of the vicuna in the puna zone of Peru. *Ecology and Management of Wild Grazing Animals in Temperate Zones.* Symposium. I.U.C.N. Warsaw, Poland.

RENNER F.G. & LOVE L.D. (1955) *Water-Yearbook of Agriculture.* U.S. Dept. of Ag. 751 p.

SMITH H.N. & RECHENTHIN C.A. (1964). Grassland restoration—the Texas brush problem. *U.S. Dept. of Ag. Soil Cons. Ser.* 17 p., 16 maps.

STODDART A. & SMITH A.D. (1955). *Range Management.* McGraw Hill Book Co. Inc. New York. 433 p.

## PART IV. EXCLOSURES

### HAROLD F HEADY

Influences of grazing animals on their habitat are studied by prevention of grazing with exclosures or by enclosing certain species but not others. Exclosures and enclosures may be used in combination and established for a short period of time or built more permanently to last a number of years. This discussion is centered on permanent exclosures in natural vegetation where grazing effects of large animals are of primary interest. Small temporary exclosures, often called cages, are used extensively in range management research (Brown, 1954; Committee, 1962; Stoddard and Smith, 1955), where emphasis is on food intake and forage utilization in relation to forage production.

Permanent exclosures provide useful areas for studies of interrelationships between natural vegetation, climate, and native animals without the influence of domestic animals; plant succession without ungulate grazing; and the influence of grazing by large animals on populations of small animals. A widespread use of exclosures has been to demonstrate that grazing has resulted in extensive vegetation and habitat changes.

Factors which determine ideal exclosure size are uniformity of vegetation, soil, and slope; type and extent of edge effects along the exclosure border; and size of home range of animals that move freely through the barrier fence. Minimum size of any exclosure that remains in place more than one year should be 0·1 hectare (1/4 acre). A smaller area is subjected to undue edge effects and is all but useless for studies of natural succession of plants and animals. Its principal value is to protect experimental plot treatments from grazing rather than to contrast conditions inside and outside the exclosure.

A survey by the author of exclosures used in range management studies in the western United States determined that those smaller than 4 ha (10 acres) were generally evaluated as being too small. Many of less than 0·5 ha (1 acre) had been abandoned because of negligible or misleading results due to attraction of rodents and rabbits, to environmental changes caused by the barrier fence, and to atypical site conditions. With carefully selected uniform condition of vegetation and soil, 10 ha (25 acres) is a more reasonable minimum size and 40 ha (100 acres) may approach optimum size. A mixture of vegetation and soil types, especially where hydrological studies are included, requires still larger areas.

Permanent exclosures are not without difficulties in maintenance and interpretation of results (Stoddard and Smith, 1955). They may be sources of infestation of animals, plants, and diseases to nearby areas; epidemics within exclosures may prevent evaluation of original objectives; and fire danger or burning effects may change through build-up of organic materials. Permanent exclosures currently established will seldom enclose pristine conditions; therefore, ecosystem changes are likely to be rapid at first and continue through more than one generation of the plant species enclosed. No grazing at all may be as unnatural as heavy use by the excluded animals.

Construction of an exclosure is costly, maintenance is continuous, and replacement comes rapidly, especially in tropical areas where both wood and metal decompose rapidly. Specifications for the barrier fence depend upon the animals to be controlled, purpose of study, and the time-span of use. The most open structure to minimize effects on wind, light, temperature, and drifting snow that is consistent with the study objectives is recommended (Committee, 1962).

Whether permanent exclosures are established as small 'livestock-free areas' or 'game-free-areas' to determine grazing effects, or as large 'natural-areas' to encompass many research purposes, their principal value is to provide an experimental treatment in the study of land use. Past experience has indicated that realistic interpretations of changes within an exclosure can be obtained only with careful planning of objectives, site selection, construction, maintenance, and long-term management. Small permanent exclosures that are not parts of planned experimentation are to be avoided.

### References

BROWN D. (1954). Methods of surveying and measuring vegetation. *Commonwealth Bur. of Pasture and Field Crops. Bull.* 42. 223 p.

COMMITTEE ON RANGE RESEARCH METHODS (1962). Basic problems and techniques in range research. *Natl. Acad. Sci.—Natl. Res. Counc. Publ.* 890. 341 p.

STODDART L.A. & SMITH A.D. (1955). *Range Management.* McGraw-Hill Book Company Inc., N.Y. 433 p.

## PART V. HERBICIDES

### FRANK E EGLER

Chemical herbicides are important within the context of this guide, since they are a cheap and effective tool for the long-term management of semi-natural vegetation, when directed to the improvement of food and cover conditions for large herbivores.

Historically, the chemicals were derived from hormone-like compounds which activated growth in minute concentration but killed plants in high concentration. Since the physiologic effect on plants is not clear, it is an incorrect analogy to say they 'grow themselves to death'; death, if it occurs, is without any known antecedent increased growth. The first large-scale commercial use of herbicides was in 1946, for electric power transmission lines. The first report of woody plant research was in 1947; the first and still valid program for right-of-way vegetation management in 1949; the first report on basal-bark spraying in 1950; the first statement of the broad research opportunities in 1952; the first recommended applications for wildlife habitat in 1953 and 1957 (*see* Egler, 1957, for a list of references). Under vigorous and aggressive promotion of the chemical industry, directed toward the spraying contractors, highway departments, and utility corporations, essentially the entire amount of the then 50 million acres in right of ways in the United States has been blanket-sprayed from 1 to 30 or more times in a procedure which is now recognized as economically unsound and destructive of conservation values, especially of wildlife habitat.

The chemical herbicides used in terrestrial vegetation management for wildlife habitat should be non-persistent and non-poisonous to wildlife and man. The esters of 2, 4-D and 2,4,5 -T, as normally used, fulfill these requirements. They disintegrate by bacterial action in 6–8 weeks and are not known to cause ill effects even when taken in quantities of a spoonful or more internally (as done at industry promotion meetings, but not recommended). For two decades 2,4-D and 2,4,5-T have been on the market, and there is no evidence they will be superseded by other chemicals at this time. On the other hand, there are hundreds of different herbicides in use for cropland agriculture, and new ones appear monthly. Chemical reactions between herbicides, and between insecticides and herbicides, including 2,4-D, are not fully known; and the above statements of safety for 2,4-D and 2,4,5-T hold only when used alone and in reasonable quantities.

When these herbicides are applied to aquatic vegetation, to lakes, rivers, and other wetland habitats, they diffuse through the entire aquatic medium. They thus affect the phytoplankton and ecosystem processes dependent upon the phytoplankton. In some instances fish appear to be sensitive to 2,4-D and 2,4,5-T. These effects must be considered in aerial application of herbicides. 2,4-D is effective essentially on broad-leaved herbaceous plants: grasses are relatively resistant. 2,4,5-T and mixtures of 2,4-D and 2,4,5-T are effective on woody plants. Many different esters of the acids are in commercial use, but pronounced differences in their general effects have not appeared. The herbicides may be applied in weak water-borne solutions, especially for highly selective basal-bark spraying, stub and stump spraying, overall stem spraying, and for pouring into notches or cups in tree trunks.

Application may be selective, by knapsack spraying or nozzle attached to a power rig; or broadcast under high pressure from power rigs, or aerially from airplane or helicopter. In broadcast spraying, great care must be taken to avoid drift of liquid particles or the effects of volatility, each of which may extend several kilometers, depending on air currents. Grapes, tomatoes, roses, and cotton are usually sensitive to minute quantities in the air. Flowering and fruiting of agricultural crops and horticultural plants may be adversely affected by drift, even at great distances from the site of application.

It is important to realize that once in the plant these herbicides move 'up' distally very readily but move 'down' proximally to a very minor degree, and then only a few xylem strands and without an effective killing of the entire plant tissue. Consequently, it is easy to get a brilliant leafburn which may be little more than a temporary defoliation, although this may be all that is wished. Shoot-kill (kill of above-ground parts), followed by resprouting as for browse production, is the reaction of some species. Root-kill usually requires special selective treatment. Stump-sprouting trees will need to have the bark soaked sufficiently both to effect a 'chemical ringing' and to kill the incipient buds on the root collar. Root-suckering trees like aspen and ailanthus are often stimulated to sucker abundantly when the aboveground part is killed at that spot. It has been found effective chemically to ring the trees (do not mechanically cut) in late summer. In this manner, carbohydrates passing to the roots may carry some of the herbicide; and continued growth of the tops tends to deplete the roots.

Season of treatment is not of prime importance. Spraying tends to be ineffective if the foliage and stems are already wet with rain. Snow on the ground prevents application close to the ground as well as does a heavy herbaceous cover. For dry season or cold season deciduous forests, spraying in the dormant season allows for twig application without the waste of a simple leaf-burn.

Any program for long-term management with herbicides must take into account the mechanics of vegetation change. The greatest single deterrent to the sound use of herbicides by scientists is the widespread acceptance of the classical theory of plant succession, whereby each stage of succession is caused by, and dependent upon, the preceding stage, and in turn is the cause that leads to the next stage. This notion has been accepted despite the fact that there is not a single research paper that supports this cause-and-effect relationship of a series of stages. To the contrary, agricultural and horticultural practice is essentially dependent upon the absence of all 'weeds'; and forest practice has long recognized that ferns and shrubs may form tightly closed, stable communities invaded by trees only with difficulty. For vegetation management with herbicides, sound, independent observations must be made as to which species can invade which communities; whether their invasion is aided or hindered by that community; and whether the invasion is by new seedlings, by root suckers from peripheral plants, or from centuries-old stool sprouts long on the site.

In conclusion, vegetation management with herbicides is a new development with huge potential. Not only is the technique new, but inadequate knowledge of existing vegetation prevents an immediate, rational application. Each species, each vegetation type, and each site type offers its own problems as well as does each method of herbicide application. Programming should be at least on a 25-year basis. After herbicide application, observations should include the degrees of leaf-kill, of shoot-kill, of root-kill; the nature of the post-spraying plant community; and the stability or resistance-to-change of that community. The desirability of this type of management is revealed by the Ten Mile River situation (Pound and Egler, 1953) where low, stable, wildlife-desirable vegetation has existed in a forest region without one dollar being spent upon it since 1936; and, in 1967, there was still no evidence of invading brush that would soon require treatment.

## References

EGLER, FRANK E. (1954). Vegetation management for right of ways and road-sides. *Smithsonian Rept.* 1953, 299–322.

EGLER, FRANK E. (1957). Selected list of publications concerning the American Museum system of right of way vegetation management. *Am. Mus. Natl. Hist. Comm. for Right of ways, Rel.* 5. First page, June 1954. Second page, Feb. 1957. An unannotated list of 63 entries. Aton Forest, Norfolk, Conn.

EGLER, FRANK E. (1961). *Herbicides. Sixty questions and answers concerning roadside and right of way vegetation management.* Litchfield, Conn.: Litchfield Hills Audubon Soc.

EGLER, FRANK E. (in press). *Right of way Vegetation Management.* Smithsonian Inst. Office of Ecol.

KENFIELD, WARREN G. (1966). *The Wild Gardener in the Wild Landscape. The art of naturalistic landscaping.* Hafner Publ. N.Y. 10003. 232 p.

MCQUILKIN W.E. & STRICKENBERG L.R. (1961). Roadside brush control with 2,4,5-T, on eastern National Forests. *U.S. For. Ser. N.E. For. Exp. Sta. Upper Darby, Pa. paper* 148. 24 p.

NIERING, WILLIAM A. (1958). Principles of sound right of way vegetation management. *Econ. Bot.* 12, 140–144.

POUND, CHARLES E. & EGLER, FRANK E. (1953). Brush control in southeastern New York: fifteen years of stable tree-less communities. *Ecol.* 34, 63–73.

Vegetation management (right of way). (1966). *Encyl. Sci. and Tech.* 14, 287–288a. McGraw-Hill, New York.

*Vegetation management for right of ways. Selective maintenance for improved wildlife habitat and scenic values.* U.S. For. Ser. Eastern Reg. (Milwaukee, Wisc. 53203) (1966). 38 p.

WISC. NATL. RESCS. COMM. OF STATE AGEN. (1967). *Selective Brush Management Program for Wisconsin Roadsides.* Wisc. Cons. Dept. Soil Cons. Serv. and Univ. Wisc. Dept. Wildl. Ecol. 16 p.

## PART VI. WOODLAND AND BRUSH REGIMES

### W J EGGELING

Woodlands and brushlands are intermediate in structure and other characters between forests and grasslands. They are composed of relatively short, woody plants growing in dense or open stands (Colman, 1953). The category embraces not only the many forms of tree savannah so typical of Africa—open mixtures of trees and shrubs usually in a matrix of tall grass (Russell, 1962)—but also any woody stand that is neither true forest nor extensive plantation, thus including smaller plantings established to provide shelter or cover.

Many types of scrub and bushy growth are the result of forest destruction and the abandonment of cultivated land; others owe their origin to excessive grazing or browsing, often evidenced by serious soil erosion. The germination of the seeds of many of the trees and shrubs is improved by heat; hence, dense thickets can result from the firing of grassy areas bearing scattered scrub. Burning or cutting often stimulates stem and stool suckering or root suckering. The value of woody vegetation for watershed cover has long been appreciated. Woodlands and brush also produce useful forage for regulated livestock and wildlife grazing; unregulated use can cause widespread degradation (Colman, 1953; Stoddart and Smith, 1955; Dasmann, 1959).

Woodlands and brush provide food and shelter for many species of herbivores—both domestic and wild—either seasonally or throughout the year. Their food may be grass and other herbage, or leaf and branch browse, or both. Mostly, such habitats meet only part of the requirements of the herbivores using them; but some animals are found only in particular types of brush, e.g. gerenuk (*Lithocranius walleri*) and dik-diks (*Rhynchotragus*) in dry Acacia bush and arid scrub in Kenya and Tanzania.

Excessive numbers of wild herbivores, often the result of the extinction by man of their natural predators followed by inadequate human control, may inhibit in part or completely the successful regeneration of woody species—so, equally, may excessive number of livestock. Red deer (*Cervus elaphus*), domestic sheep and repeated moor-burn have obliterated all tree regeneration from large areas of the Scottish Highlands. Multitudes of herded goats have done the same thing in the Mediterranean region; and elephants (*Loxodonta africana*), previously held in balance by human predation, are—again in conjunction with fire—similarly destroying both parent trees and young regeneration in national parks in East Africa.

Just as important as the destruction of tree and brush cover by herbivores and their inhibition of regeneration is the encroachment of open ranges by scrub growth, due either to an insufficiency of grazing to prevent its spread or to an excessive grazing pressure on particular components of the vegetation (notably grass) leading to a marked deterioration of existing soil conditions and the development of thicket growth. This latter situation is commonly seen in semi-arid areas subject to over-use by primitive pastoralists, especially where goats as well as sheep and cattle are involved, e.g. on the southern side of the Sahara. In eastern Uganda, lush rolling grasslands have been transferred by pastoralists within a generation into eroded tracts of thorn scrub; in western Uganda, grazing by excessive numbers of hippopotami (*Hippopotamus amphibius*) has led to a similar situation. The facility with which numerous scrub species sucker has been demonstrated in many parts of Africa where attempts to clear bush for tsetse control purposes by felling and/or intensive burning have resulted in thicker stands than before. Obviously, any study seeking to evaluate the effects of grazing animals on an existing or potential woodland or scrub habitat must require *inter alia* the setting up of exclosures.

The distribution of woodland and bush may be influenced by climatic (including altitude) and edaphic (including moisture) conditions, by biotic influences, and by fire (sometimes caused naturally, more often man-made). Frequently, the pattern of woodland and scrub owes more to the effects of human action than to climate and soil; good examples are the thicket vegetation of the Lake Albert flats adjoining the upper Nile and the regrowth deciduous woodlands of several of the New England states.

The management of woodland and brush to provide permanent shelter, grazing and browse involves both plant and animal management; it may entail also the use of fire. Much scrub and woodland is seral to climax forest, and the earlier and more open stages are usually the most important to herbivores. Management of seral communities aiming to hold a plant succession in check to perpetuate a desirable condition is notoriously difficult, especially where areas of limited extent are concerned. The maintenance of mosaic patterns in larger tracts is a more practical proposition, with the deflection or re-starting of successions by felling procedures, tree-poisoning and firing all possible solutions. The creation of woodland in range habitats to provide shelter and concomitant benefits is largely a forestry matter; in general, monocultures should be avoided and maintenance of natural diversity should be the aim. Other important factors are choice of species (including the strain best suited to the environment), positioning and site preparation. Fencing to exclude grazing animals in the early stages may be essential. Woodland management is a skilled task and needs trained staff. Controlled firing to provide fresh growth of range fodder without loss of soil fertility requires in tropical countries an assessment of the advantages and disadvantages of early as opposed to late burning, i.e. a decision whether to burn before or after the previous season's growth has dried out completely (Eggeling, 1949). It is imperative in any burning operation to have instantly available sufficient man-power to prevent the fire running out of control. Patch-burning is usually preferable to extensive firing and may be as essential for the management of herbivore ranges in open woodland in the tropics as it is for red grouse (*Lagopus lagopus*) management on heather moors in Scotland.

### References

COLMAN E.A. (1953). *Vegetation and Watershed Management.* Ronald Press Co., N.Y. 412 p.

DASMANN R.F. (1959). *Environment Conservation.* John Wiley & Sons, Inc., N.Y. 307 p.

EGGELING W.J. (revised Sangster R.G.). (1949). *Elementary Forestry; a first textbook for forest rangers.* Bailliere, Tindall & Cox, London. 307 p.

RUSSELL E.W. (editor) (1962). *The Natural Resources of East Africa.* D.A. Hawkins Ltd. (in association with East African Literature Bureau), Nairobi. 144 p.

STODDART L.A. & SMITH A.D. (1955). *Range Management* (second edition). McGraw-Hill Book Co., N.Y.

# PART VII. THE RESPONSE OF HERBIVORES
# TO VEGETATION

ALLEN KEAST

Factors governing the occurrence, distribution, and abundance of large herbivores may be divided into the physical, historical, and biotic. Species differ in the ranges of temperatures at which they can function efficiently or comfortably, and this explains latitudinal and altitudinal differences in distribution. Rainfall is important directly in providing surface water and indirectly in its influence on the vegetation. Limitations to population size imposed by low water supplies have been described by Elder (1956) and Swank (1958) with respect to mule deer and Jones, *et al.* (1957) relative to bighorn sheep in North America. The catastrophic effects of recent Kenya droughts are well known. The areas of climatic extremes—the deserts, high mountains, and the Arctic—are notoriously poor in numbers of herbivore species; and seasonal movements, dictated by the necessity to maintain minimal living conditions, are well developed in the inhabitants of these places. Soils exercise a direct influence on herbivores through the vegetation. The average daily density of grazing herbivores is, for example, four times that on red-laterite as on black chernozem soils (Petrides and Swank, 1956). Physiography, as well as being important in its influence on rainfall, drainage patterns, and vegetation, may limit distributions (e.g. the Great Rift Valley of Africa), or create habitat (vide 'crag-inhabiting' herbivores like the African Klipspringer and European chamois).

Ultimately, vegetation is the all-important factor in herbivore biology in that it provides habitat, cover, and food. Many of the structural attributes of species are associated with living in and utilizing specific vegetation zones. Fortunately, an understanding of distribution in large herbivores is enormously simplified by most species being linked to one or another of the major vegetation formations. Thus the North American herbivores fall into tundra, mountain, forest, grassland, and shrubland species (Petrides, 1960); and Australian (Keast, 1959) and African mammals (Ansell, 1960; *et al.*) may similarly be divided on the basis of the major vegetation formations occurring on those continents. Africa, with its much greater diversity of species and relatively diverse physiography and vegetational range, shows greater spatial and habitat subdivisions than the other continents. Thus, there are true marsh herbivores, dambo species, thicket species, and distinct lowland and highland grassland species.

The field worker is directed to the world vegetation map in the Oxford Atlas and more importantly detailed vegetation maps for the various continents, e.g. the U.N.E.S.C.O. map for Africa (Oxford University Press), C.S.I.R.O. map for Australia, etc.

## The description of habitat

Several schemes have been advanced for describing and formalizing habitat, one of the basic tasks of the mammal ecologist. These take into account such variables as the height of the component trees, spacing of trees, ratio of tree to shrub to grass layers, height of canopy, density of canopy and number of strata. Elton and Miller (1954) provide a general discussion of some of the variables to be considered in habitat assessment, but it is not written specifically with respect to large mammals and has only limited relevance here. The scheme of Emlen (1956) relative to avian habitats should be read for general background and for the somewhat ingenious use of symbols in describing subdivision. It is, however, cumbersome and irrelevant. The most appropriate scheme is that of Christian and Perry (1953) in which three categories of tree height and three of shrub height are introduced and symbols are provided for density and composition of dominant species. Needless to say, the mammalogist may ultimately want to introduce a descriptive scheme of his own giving appropriate emphasis to the variables that he finds to be of particular importance in the group of mammals he has under study.

### The important features of habitat

Much of the worker's energies, right from the onset of his program, will be directed at determining what the important features of the habitat are to the mammal. Much can be achieved by patient observations through field glasses and no chance should be lost to take full notes.

Included in the data that can be obtained by this means are:
(1) The minimum sized living area acceptable to the species (from amplitude of daily and seasonal movements).
(2) Degree of dependence on surface water, vide frequency of drinking, tendency to keep close to streams, etc.
(3) Responses and adaptations to normal and abnormal climatic phenomena, floods, droughts.
(4) Shelter needs (i.e. to avoid exposure, predation).
(5) Tendency to be dispersed uniformly or concentrated where there is a particular food, plant association or degree of cover. Is the distribution continuous or in the form of a mosaic?
(6) Responses to seasonal changes in vegetation, firing of the habitat.
(7) Link with particular vegetational growth stages, if any. For example, the American white-tailed deer reaches its peak of abundance when abandoned farms are at the shrub stage, e.g. when the animals can reach maximum foliage. It is well-known to ornithologists that during the transition from farm to forest the species composition may change successively; this may equally be witnessed in agricultural Europe, central Africa, and insular New Guinea.

(8) Much information of basic food and seasonal changes in diet.

(9) Needs in the way of spawning grounds.

During the above studies thought should be given to reasons for the animals being absent from other areas.

### Examples of the link between large herbivores and vegetation

The above data is basic to any management program, as are a prerequisite to understanding occurrence and numbers and to predicting what species may be introduced. The interrelationships between an animal and its habitat are complex. To show this, as well as providing examples of methods that may be used in carrying out habitat studies, reference can best be made to two outstanding African studies, that of Vesey-Fitzgerald (1960) in the Rukwa Valley, and Lamprey (1966) respecting the Tarangire Reserve in Tanzania.

(i) **Habitat preferences in animals of the Tarangire Reserve.** Reproduced as Figure 13.4 are figures for three of fourteen large herbivores, based on 3 parallel transects 8000 yards long and 1000 yards apart, carried out by pairs of observers for 20—26 days per month over a 4-year period. Counts are given in the vertical scale in the middle of each graph.

Impala obviously concentrate in the *Acacia* woodland, more specifically in what Lamprey calls the 'boundary zone' between the medium density *A. drepanolobium* zone and the open *A. tortilis* woodland, the animals finding a greater variety of food here, as well as good cover. Giraffe (*Giraffa camelopardalis*), feeding mainly on the tops of small trees, are more uniformly distributed. Hartebeest (*Alcelaphus buselaphus*) is an open grassland species with preference for riverine grassland.

(ii) **Seasonal changes in habitat utilization in the ungulates of the Rukwa Valley.** This valley, in southwestern Tanzania, consists of a valley floor containing a shallow saline lake surrounded by extensive flood plain grassland bordered by open woodlands (of Acacia and other trees). It is enclosed on two sides by high escarpments and relatively self-contained. The annual distributional and grazing cycle of the 18 herbivore species fluctuates with the alternating wet and dry seasons.

Table 13.5 summarizes the distribution of the species at different times of the year. While some are constant to a single zone, most move between

*Chapter 13*

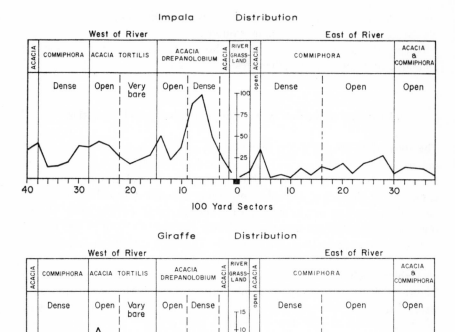

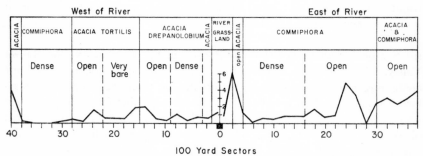

Figure 13.4. Distribution of animals on a transect in Tanganyika (redrawn from Lamprey, 1963). Vertical scale is animals per square mile.

TABLE 13.5. Seasonal habitat changes in the large herbivores of the Rukwa Valley, Tanzania (from Vesey-Fitzgerald, 1960).

| Animal species | Jan. | Feb. | Mar. | April | May | June | July | Aug. | Sept. | Oct. | Nov. | Dec. |
|---|---|---|---|---|---|---|---|---|---|---|---|---|
| Elephant | ACACIA AND ESCARPMENT WOODLANDS | | | | | | | FLOOD PLAIN | | | WOODLANDS | |
| Buffalo | | WOODLANDS—LAKESHORE AND DELTA GRASSLANDS | | | | | | FLOOD PLAIN | | | WOODLANDS | |
| Hippo | | | RIVER FRINGE AND DELTA GRASSLANDS | | WANDER WIDELY ALONG DRAINAGE CHANNELS, INTO WOODLANDS, ETC. | | | | | | | RIVER FRINGE |
| Puku | | DELTA AND LAKE SHORE GRASSLAND ALL YEAR BUT IN SEVERE DROUGHT FOLLOW RIVER UPSTREAM TO THE FLOOD PLAIN | | | | | | | | | | |
| Topi | PERIMETER GRASSLANDS | | | | LAKESHORE AND DELTA GRASSLAND | | | | VOSSIA PASTURE | | ACACIA WOODLANDS | |
| Zebra | ACACIA WOODL. | PERIMETER GRASSLANDS | | | ACACIA PARKLAND | | FLOOD PLAIN GRASSLAND | | | | ACACIA WOODLANDS | |
| Bohor Reedbuck | | | | | FLOOD PLAIN GRASSLAND | | | | | | | |
| Eland | DRY PERIMETER PLAINS | | | | DELTA GRASSLANDS AND VOSSIA PASTURE | | | | | | ACACIA WOODLANDS | |
| Giraffe and Impala | | | | | ACACIA GRASSLAND | | | | | | | |
| Warthog | | | | ACACIA GRASSLAND AND FOREST EDGE | | | | | | | | |
| Waterbuck, Duiker, Bushbuck and Steinbuck | | | | | WOODLANDS | | | | | | | |

woodlands and floor plain, depending on the degree of dryness or flooding. An interesting feature is that the movement patterns vary with the species, stressing differing ecological needs. Changes from the regular pattern occur during exceptional droughts. Topi show a marked tendency to move on to places that have recently been fired. Another interesting feature is that the heavier animals, by tramping down the vegetation, may prepasture for the lighter-bodied ones. Vesey-Fitzgerald's studies also bring out how major changes in diet accompany the seasonal movements. A large foraging area is thus necessary for the maintenance of the Rukwa fauna: if the species be artificially denied access to any part of it, they will succumb and the whole area will go to waste.

### References

ANSELL W.F.H. (1960). *Mammals of Northern Rhodesia*. Govt. Printer. Lusaka. 155 p.
CHRISTIAN C.S. & PERRY R.A. (1953). The systematic description of plant communities by the use of symbols. *Ecol.* **41**, 100–105.
ELDER J.B. (1956). Watering patterns of some desert game animals. *J. Wildl. Mgmt.* **20**, 368–378.
ELTON C.S. & MILLER R.S. (1954). The ecological survey of animal communities with a practical system of classifying habitats by structural characters. *J. Ecol.* **42**, 460–496.
EMLEN J.T. (1956). Method for describing and comparing avian habitats. *Ibis.* **98**, 565–576.
JONES F.L., FLITTNER G. & GARD R. (1957). Report on a survey of bighorn sheep in the Santa Ross Mountains, Riverside County. *Calif. Fish and Game.* **43**, 179–191.
KEAST A. (1959). Vertebrate speciation in Australia: some comparisons between birds, marsupials, and reptiles. In: *The Evolution of Living Organisms*. Melbourne Univ. Press. p. 380–407.
LAMPREY H.F. (1963). Ecological separation of the large mammal species in the Tarangire Game Reserve, Tanganyika. *E. Afr. Wildl. J.* **1**, 63–92.
PETRIDES G.A. (1960). The management of wild hoofed animals in the United States in relation to land use. In: *Ecology and Management of the Wild Grazing Animals in Temperate Zones*, 8th Tech, Meet. I.U.C.N. Warsaw. 181–202.
PETRIDES G.A. (1963). Ecological research as a basis for wildlife management in Africa. In: *Conservation of Nature and Natural Resources in Modern African States*. I.U.C.N. Morges, Switzerland, 284–293.
SWANK W.G. (1958). The mule deer in Arizona chapparal. *Ariz. Game and Fish Dept. Wildl. Bull.* 1. 154 p.
VESEY-FITZGERALD D.F. (1960). Grazing succession among East African game animals. *J. Maml.* **41**, 161–172.
VESEY-FITZGERALD D.F. (1965). The utilization of natural pastures by wild animals in the Rukwa Valley, Tanganyika. *W. Afr. Wildl. J.* **3**, 38–48.

# PART VIII. CONTROLLED GRAZING

## KENNETH W PARKER

Controlled grazing of domestic livestock entails skillful manipulation of grazing use to provide for growth requirements of forage plants, prevent soil deterioration, and allow sustained maximum animal production. Principles include: (1) proper balancing of numbers of animals with the forage crop; (2) proper kind and class of livestock as determined by range suitability to provide for maximum animal production; (3) season and sequence of use adjusted to the physiological needs of the important forage plants and the ecological requirements of the plant community; and (4) proper distribution of livestock so as to attain full use of the range. On deteriorated range, application of these principles may be of little avail unless preceded by effective cultural improvement programs such as seeding, noxious plant control or fertilization.

In North America the various kinds of large tame and wild herbivores differ greatly in their grazing habits. In general, cattle and elk show a marked preference for grass, and sheep and antelope favor broad-leaved herbs or shrubs or forbs and browse. On the other hand, goats and deer make the greatest use of browse but, in season, consume both grasses and forbs. Consequently, cattle compete with elk the most, and sheep and goats most with deer. In many parts of the world, integration of use by domestic livestock with the large wild herbivores is an ever-mounting problem. Its solution lies in providing for and balancing the needs of all kinds of animals with forage production. However, this may be difficult, especially in the case of big game animals whose populations must be controlled by hunter harvest.

All ranges are best maintained in satisfactory condition if grazed at the time when the physiological shock to the important forage plants is at a minimum. With grasses the most critical periods are usually at the start of growth when the plants are dependent on stored food reserves and during flower stalk formation and seed ripening. However, under practical livestock range management it is not always possible to graze in complete accordance with the plant growth requirements because grazing during the 'green feed' period is also optimum for animal production. The next best thing is to adjust the intensity and seasons of use so that the effects of grazing are not beyond the physiological tolerance of the important forage plants.

Inventory of the range as to species and quantity of forage plants, stages of plant succession, site productivity, soil stability, suitability for grazing use, and existing or needed physical improvements such as fencing and water development is essential for planning the management. Such information allows adoption of grazing systems and other management techniques that will best meet the requirements of vegetation and animals. Periodic observations on range condition and trend are essential to properly assess the effectiveness of management.

Some ranges can be grazed seasonlong or yearlong by domestic livestock, provided the degree of use on it is not excessive. However, even under low stocking rates, some plants are closely grazed repeatedly; and periodic rest is needed to restore vigor and allow satisfactory production of forage and seed for establishment of new plants. This can be accomplished by dividing the range into two or more units and applying some system of rotation, deferred, or rest-rotation grazing. This alternate protection and use in specified sequences is designed to favor the most desirable species.

Such management practices will normally allow satisfactory range maintenance or improvement; however, if severe deterioration has occurred, seeding or plant control measures may be necessary. Also, manipulation of vegetation may be desirable to improve forage production, particularly for big game. Thinning stands of timber, creation of forest openings, prescribed burning, seeding or planting, fertilization, etc. may all be used effectively to restore deteriorated habitats or improve those naturally unproductive.

### References

DAUBENMIRE R.F. (1947). *Plants and Environment. A textbook of plant autecology.* John Wiley & Sons, Inc., New York. 1–424.

DIETZ D.R. (1965). Deer nutrition research in range management. *Thirtieth N. Amer. Wildl. and Natur. Res. Conf. Trans.,* 274–285.

DRISCOLL R.S. (1967). Managing Public Rangelands: effective livestock grazing practices and systems for national forests and national grasslands. *U.S. Dept. Agr., Agr. Inf. Bull.* 315. 30 p.

DYKSTERHUIS E.J. (1949). Condition and management of range land based on quantitative ecology. *J. Range Mgmt.* **2**, 104–115.

ELLISON L. (1960). Influence of grazing on plant succession of rangelands. *The Bot. Rev.* **26**, 1–78.

HALLS L.K., HUGHES R.H., RUMMELL R.S. & SOUTHWELL B.L. (1964). Forage and cattle management in longleaf-slash pine forests. *U.S. Dept. Ag. Farmers' Bull.* 2199. 25 p.

HANSON H.C. & CHURCHILL E.D. (1961). *The Plant Community*. Reinhold Pub. Corp., N.Y. 218 p.

JULANDER O. (1955) Deer and cattle range relations in Utah. *Forest Sci.* 1, 130–139.

KENDEIGH S.C. (1961). *Animal Ecology*. Prentice-Hall, Inc., Englewood Cliffs, N.J. 468 p.

PARKER K.W. (1960). Principles of grazing management as related to vegetation condition and soil stability. Forest and Range Watersheds. *Fifth World Forest. Congr. Proc.* 3, 1708–1714.

RENNER F.G. & ALLRED B.W. (1962). Classifying rangeland for conservation planning. *U.S. Dept. Ag. Ag. Handbook* 235. 48 p.

SAMPSON A.W. (1952). *Range Management Principles and Practices*. John Wiley & Sons, Inc., New York, Chapman & Hall, Ltd., London. 570 p.

STODDART L.A. & SMITH A.D. (1955). *Range Management*. 2nd Ed. McGraw-Hill Book Co., Inc. Toronto. 433 p.

# PART IX. INSECTS AS COMPETITIVE HERBIVORES

## C B HUFFAKER

### Significance of insects as competing herbivores

Phytophagous insects contribute an ever-present element of cohabitation and, to some extent, direct competition with the large herbivores, although, of course, the intensity of such competition may be either heavy or light. The herbivorous action of insects is usually far less obvious than that by large herbivores, unless certain species reach epidemic abundance. This is because many insects are more selective in their grazing than the large herbivores; and, in a heterogeneous complex of plant species, the result will be that certain forage species are affected more or less so, relative to others, at a given time and place, while at other times and places the contrary will prevail. Thus, there may be a greater tendency to alter the precise composition of the total yield available to the large herbivores than to substantially reduce that total yield. Nevertheless, on given stands of vegetation, one or a few particular species of plants make up most of the forage available. Substantial destruction of the dominants by an insect having sufficient dietary latitude, or by several species more dietetically limited, can, and often does, greatly deplete the forage available to ungulates and other large herbivores. In this pattern of selective destruction of certain species only, there is a substantial capacity of the range vegetation to compensate through increased growth of the undamaged species that may fill in the void. The extent to which selectively-feeding phytophagous insects might thus influence plant composition if it were not for the action of their own enemies on them is difficult to comprehend. But we are not concerned with this latter aspect here, except to note that occasional or periodic 'escape' of such phytophagous insects from the controlling effect of their own enemies is commonly an explanation of a newly achieved outbreak status.

High utilization of forage by insects has been noted by man since earliest recorded history. Biblical records refer to the vast hordes of locusts that swarmed the skies of the Near East then as they do still today. In its swarm phases the migratory locust *Schistocera gregaria* is decisively polyphagous and destroys nearly any kind of green food at hand. The writer saw vast stretches of semi-dry grazing lands in Morocco swept clean of vegetation by this species in 1954. A key to control or prevention of massive outbreaks of the swam phase is linked with a thorough knowledge of its solitary-phase biology, habitats and preferred food plants (Gunn, 1960). Pasture grubs (various species of Scarabeidae) in parts of New Zealand often become so abundant that the total biomass of these grubs (competitively reducing large herbivore productivity) beneath the surface of the turf is much greater than the biomass of sheep that can be carried there (Given, 1966). Wolcott (1937) estimated that in normal pastures of good vegetation in New York where cows averaged 1 per 8 acres, the dry weight of cow per acre was roughly only twice that of the dry weight of the insects living there; and, if the other invertebrates and small vertebrates were included, the dry weights of cow and small animals were about equal. He found that often the insects ate more of the forage than the cows did. Pringle (1955) and Arnott (1957) reported destruction of sagebrush *Artemisia tridentata* by the beetle *Trirhabda pilosa* in the northwestern United States of America and a consequent increase in forage grasses.

That relatively host-specific phytophagous insects may alter vegetative composition to the *advantage* rather than the *detriment* of large herbivores while competing with them to a degree has also been demonstrated many times by programs of biological control of particularly weedy components of range and pasture vegetation. This is the other, equally important side of the question. C. J. Davis (personal communication) has recently summarized the cases of partial, substantial, or complete biological control of respective weeds in various countries of the world. There have been 23 cases of substantial or complete control: 14 of these were complete successes, with very few complete failures. The cases of virtually complete change of vegetation from a status of almost no grazing value whatever to restoration to full grazing capacity by the introduction of the moth *Cactoblastis cactorum* into Queensland for control of prickly pear and of the beetle *Chrysolina quadrigemina* into California for control of St. John's wort or Klamath weed are classical examples (Dodd, 1940; Huffaker and Kennett, 1959).

The fundamental lesson here is the potential exhibited by such selective insect grazers in altering the complex of vegetation. With many of the cases of successful biological control of pernicious weedy species, it has been necessary to try many different kinds of insects before the most effective one(s) was found; and the best species in one situation is not necessarily the best in another.

Abundant evidence suggests that such enemies have often comparable roles in the control of the same host species in the regions where both host plant and enemy evolved and are endemic. The evidence refutes the common suggestion that only exotic species can have such striking and fundamental impact. It is especially pertinent that in natural endemic situations effective host-specific insects would be expected to be relatively more scarce—i.e., proportionate in abundance to the abundance of the host plant itself under the degree of control that pertains (Huffaker, 1957, 1962).

## Methods of studying the impact of insects

This brings us to the question of proper methods of study, for we see that it is particularly necessary to study the low-density endemic situations. Good methods for doing this have not been developed.

In any appraisal of the role of insects as competitors, or as 'allies' of large herbivores as here discussed, one should not simply undertake to measure the direct destruction of the vegetation for a specific interval of time, as if that would give the answer. The indirect role of selective phytophagous insects in altering plant composition by reversing the competitive advantage of one species over another may be important far beyond the indicated measure of its consumption of total forage, as if that alone is a clear-cut denial of the forage by the same amount, to a large herbivore (Wilson, 1964 and Huffaker, 1957). Consuming insects may by such action be a vital force in reducing the amount of total forage they will consume subsequently and in increasing, to a corresponding degree, the amount subsequently available to large herbivores. No one has yet made a detailed study of the compensatory aspects involved regarding potentially important insects associated with a major type of range or pasture vegetation. No one has even tried, over a prolonged period of time, to exclude all the phytophagous insects as a group from plots in such vegetation. This would be a first step and, in conjunction therewith, if means could be devised to selectively exclude certain species and not others, further insights would be gained. It is strange that much effort has been devoted to use of exclosures to measure the influence of large herbivores and smaller mammals on plant growth and composition, but relatively nothing has been done to similarly measure the influence of the much more selective grazers among the insects. We note, however, that the direct short-term competitive impact of such insects as grass grubs, wireworms and certain others of direct competitive importance have, however, been so studied (Osborn, 1939; Lilly, 1956; Kuhnelt, 1963).

Other quantitative studies on the impact of certain insects on pasture or range vegetation have for the most part been centered on control of single weedy plant species and the pattern of replacement vegetation ensuing. Huffaker (1951), Huffaker and Kennett (1959), Clark (1953) and Clark and Clark (1952) used quadrat analyses or line transects to measure the influences associated with control of St. John's wort by the introduced beetle (*see* above). Line transects and extensive use of quadrants are being used in California for appraisal of the influence of still other insects introduced for control of tansy ragwort, *Senecio jacobaea*, puncture vine, *Tribulus terrestris*, and yellowstar thistle, *Centaurea solstitialis*.

Unfortunately, use of the powerful tools of regression and life table analysis of Morris (1963) and Nielson and Morris (1964), or the 'key-factor' approach of Varley and Gradwell (1960) and Huffaker and Kennett (1966) for assaying the impact of various mortality factors on insect pest species have not been developed or adapted for use concerning control factors operating on plant populations. Erratic germination, indistinct generations, and difficulties in assaying plant population abundance are among the problems presented. Until such techniques or others that will serve the purpose are developed, the necessary crucial analysis of the role of insects in determining plant composition in the sense of regulating the abundance of certain key component species cannot be made. Until this is done, efforts to estimate the competition large herbivores face from phytophagous insects, or the degree to which they benefit by it, will give only a very inadequate picture, except in the case of non-selective insect species.

## References

ARNOTT D.A. (1957). Occurence of *Trirhabda pilosa* Blake (Coleoptera: Chrysomelidae) on sagebrush in British Columbia, with notes on life history. *Proc. Ent. Soc. Br. Columbia* **53**, 14–15.

BROWN D. (1954). Methods of surveying and measuring vegetation. *Bull.* 42. *Commonwealth Agricultural Bureaux, England.* 223 p.

CAIN S.A. & CASTRO G.M. DE O. (1959). *Manual of Vegetation Analysis.* Harper & Bros., N.Y. 325 p.

CLARK L.R. (1953). The ecology of *Chrysomela gemellata* Rossi and *C. hyperici* Forst. and their effect on St. Johnswort in Bright District, Victoria. *Aust. J. Zool.* **1**, 1–69.

CLARK L.R. & CLARK N. (1952). A study of the effect of *Chrysomela hyperici* Forst. on St. Johnswort in the Mannus Valley, N.S.W. *Aust. J. Ag. Res.* **3**, 29–59.

DODD A.P. (1940). *The Biological Campaign against Prickly Pear.* Commonwealth Prickly Pear Board (Australia), Brisbane, Australia, 177 p.

GIVEN B.B. (1966). Biological control of weeds and insect pests in New Zealand. Natural Enemies in the Pacific Area (Biological Control). *Mushi* **39**, Suppl., 131 p.

GRIEG-SMITH P. (1964). *Quantitative Plant Ecology.* 2nd Ed. Buttersworth & Co., London. 256 p.

GUNN D.L. (1960). The biological background of locust control. *Ann. Rev. Ent.* **5**, 279–300.

HUFFAKER C.B. (1957). Fundamentals of biological control of weeds. *Hilgardia* **27**, 101–157.

HUFFAKER C.B. (1962). Some concepts on the ecological basis of biological control of weeds. *Can. Ent.* **94**, 507–514.

HUFFAKER C.B. & KENNETT C.E. (1959). A ten-year study of vegetational changes associated with biological control of Klamath Weed. *J. Range Mgmt.* **12**, 69–82.

HUFFAKER C.B. & KENNETT C.E. (1966). The biological control of olive scale *Parlatoria oleae* (Colvee), through the compensatory action of two introduced parasites. *Hilgardia* **37**, 283–335.

KUHNELT W. (1963). Soil-inhabiting arthropoda. *Ann. Rev. Ent.* **8**, 115–136.

LILLY J.H. (1956). Soil insects and their control. *Ann. Rev. Ent.* **1**, 203–222.

MORRIS R.F. (ed.) (1963). The dynamics of epidemic spruce budworm populations. *Mem. Ent. Soc. Can.* **31**. 332 p.

NEILSON M.M. & MORRIS R.F. (1964). The regulation of European spruce sawfly numbers in the maritime provinces of Canada from 1937 to 1963. *Can. Ent.* **96**, 773–784.

OSBORN H. (1939). *Meadow and Pasture Insects.* Educator's Press. Columbus, Ohio. 288 p.

PRINGLE W.L. (1955). A new look at sage brush control. Presented at the annual meeting of the Pacific Northwest Sect. Am. Soc. Range Mgmt., Yakima, Wash.

VARLEY G.C. & GRADWELL G.R. (1960). Key factors in population studies. *J. Anim. Ecol.* **29**, 399–401.

WILSON F. (1964). The biological control of weeds. *Ann. Rev. Ent.* **9**, 225–244.

WOLCOTT G.N. (1937). An animal census of two pastures in northern New York. *Ecol. Monog.* **7**, 1–90.

## PART X. CONTROLLED BURNING TO
## IMPROVE RANGELAND

### HAROLD H BISWELL

Control burning is an effective and economical means to improve ranges, especially where shrubs are involved. Shrubs sometimes become old and unpalatable or dense and tall, being out of reach of large herbivores. Burning may be used to thin the shrubs and to bring about sprouting and new seedlings, both of high palatability and nutritional value. Very often control burning must be used in conjunction with other means for best results, such as smashing with a bulldozer ahead of burning, and followed by herbicides to remove certain unpalatable species. The techniques will vary from one brush cover type to another and also with the habitat requirements of the herbivores. For example, cattle may do best on ranges heavily burned and relatively free from shrubs where grass grows instead. On the other hand, sheep profit from a considerable abundance of desirable browse species; and thus meat production may be highest in open stands of shrubs.

Deer show marked preference for browse; and burning, along with other treatments, should be utilized to maintain shrubs in a healthy nutritious condition. Usually control burning for sheep and deer is different to the extent that sheep require more open ranges so they can move from one burned area to another without difficulty.

Studies of control burning in chamise brush in Lake County, California, can be cited as an example of the use of fire in game range improvement. In general, the chamise brushlands in this area comprise two cover types: one in which chamise (*Adenostoma fasciculatum*) predominates on the south-facing exposures, and one containing a mixture of broadleaf shrubs and trees on the north-facing exposures. Drinking water is sufficient in the ravines. The shrubs may become very dense and unpalatable to deer. Such stands have little herbaceous vegetation in the understory for forage when the shrubs are dormant in winter.

Manipulation can be accomplished by burning followed by reseeding to herbaceous vegetation. The deer can then browse the sprouts and seedlings as they grow back. Another means is to burn in spots, then reseed and employ chemicals for a complete conversion to grass, thus creating 'edge' around which the deer can browse.

When control burning is contemplated, a permit should be obtained from the proper authorities and neighbors should be notified of such intent. Usually fire lanes 10 to 25 feet wide must be constructed around the area to be burned; however, there are certain exceptions. When the brush cover is predominately chamise on south exposures and mixed chaparral on north exposures, spring burning may be done without excessive risk and without wide firebreaks. The south-facing chamise slopes can be burned on days of relatively low humidity and with proper wind velocity and direction from February through May, for then the north-facing slopes of mixed chaparral are not very likely to burn. On quiet days, fires lit at the bottom of chamise slopes usually go out at the ridge top and ordinarily no more area will be burned than the length of slope by the distance fired along the bottom. One or two men equipped with flame throwers can usually do the job. However, throughout this period there will be many days when it is too moist to burn and others when it is too dry to use fire safely.

Where chamise brush occupies all exposures and is of uniform density, the risk of using fire is greater than where the type of brush varies with exposure. With a uniform brush cover, the fire is more likely to spread because conditions for burning are more uniform also. Where grass borders the brush, burning when the grass is green adds an element of safety.

The extent to which a brush area should be opened depends very largely on the deer population present. When a square mile has less than ten deer a half dozen scattered, opened areas, each of about five acres, may be sufficient. Where the population is greater than ten per square mile, a correspondingly larger number of acres should be opened. Approximately 30% of a brush area should be left unburned in well-distributed spots as cover for the deer.

All burned areas of dense brush should be reseeded to adapted herbaceous forage plants before the first fall rains. After an area is opened and the proper balance attained between dense brush and openings, burning should be done carefully or the dense brush cover may be destroyed and an area opened too much.

Occasionally, ranchers and sportsmen get together for control burning operations when the vegetation is dry and there is danger of fire escape. In this case a date is set for burning and careful plans made for the fire. Such burns are of a size that can be completed in one day.

Studies have shown that in burned brushlands the number of deer, the ovulation rate, and the weight of bucks at harvest time are all considerably greater than in unburned areas.

### References

Biswell H., Taber R.D., Hedrick D.W. & Schultz A.M. (1952). Management of chamise brushlands for game in the north coast region of California. *Calif. Fish and Game.* **38**, 453–484.

Biswell H.H. (1963). Research in wildlife fire ecology in California. *Proc. Second Tall Timbers Fire Ecol. Conf. Tallahassee, Fla.* p. 63–97.

Komarek R. (1963). Fire and the changing wildlife habitat. *Proc. Second Tall Timbers Fire Ecol. Conf. Tallahassee, Fla.* p. 35–43.

Miller H.A. (1963). Use of fire in wildlife management. *Proc. Second Tall Timbers Fire Ecol. Conf. Tallahassee, Fla.* p. 19–30.

Spencer D.L. & Hakala J.B. (1964). Moose and Fire on the Kensi. *Proc. Third Tall Timbers Fire Ecol. Conf. Tallahassee, Fla.* p. 11–33.

Stoddard H.L., Sr. (1962). Some techniques of controlled burning in the deep Southeast. *Proc. First Tall Timbers Fire Ecol. Conf. Tallahassee, Fla.* p. 133–144.

## PART XI. THE USE OF FIRE AS A TOOL IN WILDLIFE MANAGEMENT IN THE KRUGER NATIONAL PARK

### U de V Pienaar

The Kruger National Park forms part of the great savanna eco-system of tropical and sub-tropical Africa. In this semi-arid, summer rainfall area, the undisturbed virgin veld is typically perennial, tufted or bunch grassland, studded with various woody plants in the shape of trees, shrubs, and bushes. The amount of bush and trees in relation to grass varies greatly, from open veld in which woody growth is all but absent, through parkland to woodland and finally dense bush or thicket, in which grasses are quite unimportant. The predominant 'bushveld' aspect represents a stage in plant succession where the two competing communities of grasses and associated forbs on the one hand and woody shrubs and trees have met, and where the former will, as a rule, eventually give way to the latter in the absence of fire.

Although the Kruger National Park has been in existence for more than 60 years and has been protected as such from destructive influences as far as possible, much of its vegetation is not 'climax' in the edaphic or climatic sense, but is derived or fire induced.

The physiognomic aspect of the vegetational milieu is subdivided into some six major veld types, each differing from the other in floristic composition and usually associated with circumscribed areas of particular soil structure and divergent climatic regime. The different veld types in the Kruger National Park are today maintained in their present state largely by the delicately balanced interaction of climate, soils, vegetation, animals and fire.

Apart from the conservation and stabilizing of water resources, the control of unwanted vegetation with the aid of herbicides, the culling of excessive numbers of herbivorous animals, and the control of competitive herbivores (small animals and insects), the judicious use of fire is about the only practical means by which the wildlife manager can manipulate the vegetational component of the particular eco-system to suit the needs of the animal life which is required to derive maximum benefit from it.

Prior to 1950, veld-burning had been left much to the discretion and whims of the officer in charge of a particular section of the Park. There was no generally applicable policy, although it was customary to employ 'early burns (i.e. in fall) to attract game to tourist roads in winter and to burn all old, matted grass in late winter, thus producing a fresh flush of young grass in spring, as well as controlling excessive numbers of external parasites (particularly ticks).

Although accidental wild fires caused by lightning, poachers, etc. were extinguished and prevented from spreading if possible, there was no organized fire-fighting system.

Since 1950, considerable strides have been made in establishing an efficient system of fire-breaks throughout the Park. A provisional policy of rotational veld-burning was adopted in 1954 and is still in operation today, although it has been modified several times. Scientific personnel was recruited and an extensive research program was initiated, including all aspects of veld-burning. This necessitated, among others, the laying out of experimental veld-burning plots in the different vegetation types to measure and interpret the effect of veld-burning at different times of the year on the various components of the vegetation and the animal life utilizing them.

Although these experiments have not been concluded, in some plots obvious trends are already becoming evident; and the control plots provide valuable information on vegetational succession in the complete absence of fire.

The fire-break system of the Kruger National Park consists of a well-maintained road network which is graded early in the dry season to a minimum width of 24 feet. In such areas where it is impossible to grade the breaks effectively, such as along border rivers and in rocky or mountainous regions, the strips are hoed manually.

The efficiency of the boundary fire-breaks is usually further improved by burning a strip along their length late in summer or early fall. This effectively prevents the entry of wild fires from the adjoining Portuguese territory or local African settlements which border on the reserve.

The fire-break network divides the whole Park into a series of blocks varying in size from about 15 to 50 sq miles. The vast expanse of the Park (7340 sq miles) has made it desirable to set as an ultimate aim blocks of not more than 25 sq miles in extent. In areas with heavy game concentrations where the object is to establish a system of rotational grazing through veld-burning, care should be exercised not to cut the blocks too small, as this might lead to excessive utilization of the burned areas and trampling during the growth season. In such situations, adjoining blocks may, of course, be burned simultaneously.

On the other hand, in sparsely-populated regions, where the object is better year-round utilization of undergrazed areas, the blocks should not be too large. This also applies to areas where heavy grazing pressure may actually have a beneficial effect on the coarse grass stratum—i.e. the sour veld of the higher rainfall areas.

It is highly desirable that fire-breaks should be so designed that they run in straight lines wherever possible, but not at the cost of cutting obliquely through adjacent dissimilar vegetational regions. In such instances, fire-breaks should always be constructed to separate areas of different floristic composition or aspect.

Before allocating individual blocks to particular veld-burning treatments, a thorough knowledge of game habitat selection, food preferences, seasonal grazing grounds of migrating species and other important aspects such as unique floristic regions, catchment areas, springs and swamps, etc. is of prime importance.

Although the grass cover of most of the Kruger Park is what may generally be referred to as 'sweet veld,' i.e. grazing which retains much of its palatibility and nutritive value even during the dry season, it does not all receive the same veld-burning treatment.

Large areas are not burned at all and are strictly protected from fire, i.e. mountainous regions; important catchment areas: riverine strips or regions around permanent water supplies; swampland or vleis; abused areas where there are signs of erosion or which have suffered severely from drought or overgrazing; unique floristic areas which are preserved as botanical reserves within the Park, regardless of their value as game habitats; and areas with an inherently sparse and vulnerable grass cover, such as on some shallow, gravelly soils.

At present the remainder of the Park, apart from the sour veld regions around Pretoriuskop, is subjected, for the purpose of grazing control, to a system of triennial rotational burning in spring, following the first substantial rains (a precipitation of at least 2 inches is essential). The sour veld with its tall grass sward is burned biennially, and here half the area is treated in spring and half in early fall. These treatments are switched around every six years.

Before any block which is due for burning during a particular season is subjected to the treatment, it is inspected by a team of scientific personnel in order to ascertain whether it is fit for burning or not. In such cases where, by accident, a block had been completely burned out of rotation, it is not burned again during the subsequent season but is allowed to rest until it again comes into rotation during the following cycle. If only a portion of a block is lost through accidental fire out of rotation, the rest is burned during a suitable time in the same year and the block is then treated as in the former case.

It has been found necessary to limit the rotational burning of blocks in regions with dense concentrations of migrating game, such as wildebeest and zebra, except where there is a tendency to selective utilization of winter grazing grounds (which is usually not the case).

In all regions in the Park, as elsewhere, where bush encroachment has assumed serious proportions, only very hot fires immediately prior to the first spring rains can be successfully employed in controlling scrub growth, by burning the coppice stands back periodically.

A treatment of this nature is a bad risk, however, in a national park such as Kruger, where the rainfall is erratic. This is particularly true in fenced areas, where the game is denied free access to grazing in outlying regions, should the rains fail in the park and no growth be stimulated on the burned areas. Very serious overgrazing can result from such burns and the resultant bush encroachment may be worse than that which it was intended to control by the late winter or early spring burn.

Whereas this form of fire usage is still the practice in many parts of Africa, and it is doubtless instrumental in maintaining the open savanna aspect of these regions, it would be advisable to control undesirable bush encroachment in areas where they are subjected to periodic droughts by mechanical means or through use of the modern, highly effective arboricides rather than by applying late burns. This is also true for areas where bush and thickets have already been firmly established over preceding years through the indiscriminate use of fire or prolonged overgrazing.

Certain fire-resistant and unpalatable grasses, such as *Bothriochloa insculpta*, may be stimulated and seriously infest good pasture in situations where selective grazing is continually encouraged by too frequent burning. Fortunately, it appears that such grasses have a limited ability to compete successfully with the 'climax' grass species if such infested veld is protected from fire for a period of 5 years or more. This can obviously only be done on a limited scale and is best accomplished by withdrawing one block at a time from the rotational burning program.

In such areas where it is advisable to limit the disastrous effect of late burns on the woody vegetation to an absolute minimum, and where serious erosion may follow on heavy spring downpours over denuded veld, absolute protection from fire would be the ideal. Because this is seldom possible, and in view of situations arising where the thick swards of old, unburned grass in such areas actually create very dangerous fire hazards, such areas in the Park are burned during an opportune time when only a singeing effect is created. This is usually accomplished in late summer; and burns are, if necessary, completed at night. All dry, combustible material is eliminated in this way, without damage to tree and shrub strata; and some green growth of grass may even be stimulated before the onset of winter.

A treatment of such nature every second year is usually sufficient to protect such an area from serious fire-damage; but in undergrazed areas with particularly dense stands of grass, and even in border areas where poachers fire the veld at every opportunity during the dry season, it may be necessary to resort to annual early burns of this nature. This has the added advantage of preventing game from straying out of the Park during the dry season. The only danger involved in such a scheme is that it may lead to very heavy grazing pressure on such burned areas during the dry season, particularly in the vicinity of permanent water.

Once a veld-management policy involving the use of fire has been established for a particular region, accidental fires should be stringently combated by trained and well-equipped fire-fighting crews whenever they arise. The methods employed for this purpose vary according to the topography and vegetational aspect of the region involved, but in the Kruger Park veld fire control is conducted very much along the lines described by Gammon (1962) for Rhodesian conditions.

As pointed out above, the present veld-burning policy operating in the Kruger Park is by no means final, and will be subject to changes indicated by the results of current veld-burning research, as well as unforeseen future demands.

It is deemed necessary, in any event, that all interim veld-management policies involving the use of fire should be substantiated by a proper series of controlled veld-burning experiments. Three basic treatments should be incorporated, e.g. an early burn (late summer or early fall), late winter burn, and a spring burn after substantial rains had fallen.

The application of the treatments could follow an annual as well as a biennial, triennial or even longer, rotational pattern, as the occasion demands, and should not be limited to a definite month or period. In every case the block should only be fired during a time when the desired effect is most likely to be created. The late winter burn, for instance, should only be applied when the veld is really dry and the resultant fire is likely to be very hot. Only in this manner will the great natural conflagrations that have swept through the savanna regions of Africa every 2nd or 3rd year since time immemorial be truly simulated. Conversely, it will be of no avail to apply the spring burn before sufficient rain has fallen; and this may be delayed for several months. The size of the experimental plots should preferably be large enough to prevent an unnatural attraction and concentration of game following the treatments. If this is not the case, the very heavy grazing and trampling may obscure completely the effect of the fire and the experiment will lose its purpose. For this reason the experimental burning areas should not be established close to permanent water supplies. The original botanical surveys and the regular test surveys to follow should be conducted according to the method most suitable for the particular vegetational type.

Where a region subjected to veld-burning is composed of vegetational types of diverse nature, a series of experimental plots should be established in every veld type. For proper statistical analyses of the results, each experimental plot should have several replications in the same veld type. It is also desirable to have several control plots (which may be much smaller than the areas burned) to minimize the danger of losing valuable data should one be accidentally fired.

## References

BRYNARD A.M. (1965). The influence of veld-burning on the vegetation and game of the Kruger National Park. In *Ecological Studies in Southern Africa*. W. Junk, Den Haag.

DAVIDSON R.L. *et al*. (1961). Calibration of the belt transect method in Combretum woodland in the Kruger National Park. *Koedoe*. **4**, 31–44.

GAMMON D.M. (1962). Veld fire control. *Rhodesia Ag. J.* **59**, 177–191.

KENNAN T.C.D. (1961). Veld management in the farming areas of S. Rhodesia with special reference to veld-burning. Paper read at the Wildlife Conference Course, University College of Rhodesia and Nyasaland, 19th–26th May, 1961.

MITCHELL B.L. (1961). Ecological aspects of game control measures in African wilderness and forested areas. *Kirkia*. **1**, 120–128.

PHILLIPS J.F.V. (1930). Fire: its influence on biotic communities and physical factors in South and East Africa. *S. Afr. J. Sci.* **27**, 352–367.

TIDMARSH C.E.M. & HAVENGA C.M. (1955). The wheel-point method of survey and measurement of semi-open grasslands and Karoo vegetation in South Africa. *Bot. Surv. of S.A. Memoir*. **29**, 49 p.

# 14

# Epizootiology

## B C JANSEN

### Introduction

The investigation of the different manifestations of diseases among animals requires a specialized knowledge of the diseases concerned. The agents responsible for diseases in domestic and wild animals are, for the large part, identical. There are few, if any, diseases occurring specifically in wild herbivores to which domestic herbivores are not susceptible, with the possible exception of certain host specific internal and external parasites. On the other hand, there are certain well-known diseases of domestic animals to which wild animals are not known to be susceptible or of which wild animals can harbour without showing any clinical manifestations.

Hence the diseases occurring among domestic animals, which have been studied for many years and are well documented, serve as an excellent basis for the study of disease in wild animals, our knowledge of which is sadly lacking in many respects. It seems most essential that this discrepancy be removed in due course.

In investigating diseases, whether in wild or tame animals, the expert knowledge of veterinarians should be depended upon. For the correct evaluation and diagnosis of disease, as well as for the study of epizootics, careful observation and recording of all relevant phenomena and the submission of appropriate specimens for laboratory examination are of inestimable value.

In studying the epizootiology of diseases in wild animals certain peculiar features must be taken into consideration. Examples of such peculiarities are the following:

1. Wildlife is subject to uncontrolled movement and consequently has unlimited opportunities for spreading disease through its habitat. Restricting the movements of animals is a very potent means of controlling many diseases. In the case of epizootics in domestic stock this usually presents no problems. Where wild animals are involved it may be extremely difficult, if not impossible, to limit the spread of highly

contagious diseases such as rinderpest and foot and mouth disease. Game animals, particularly the gregarious species, have a tendency to congregate in certain favorable localities. If left undisturbed, this may cause serious overstocking to the detriment of the health of the animals and of their food and water supplies. This in turn leads to periodic or seasonal mass migrations and possible spread of disease.

2. Under certain conditions animals not susceptible to a particular disease may assist in spreading such a disease. An example of this process is the vulture engorging itself on the carcass of an animal dead of anthrax and regurgitating the infected material at a remote locality. Birds have been incriminated as being responsible for the spread of foot and mouth disease.

3. Predators usually prey on animals weakened by disease, thus removing some evidence of the presence of such disease. This emphasizes the necessity for constant vigilance on the part of those responsible for observations on game animals. In a certain area of the Kruger National Park it had been noticed for a number of years that about 50% of blue wildebeest calves were lost during the first three months of calfhood. This was attributed to predation by carnivores until an appreciable increase in the losses during one season led to an investigation. Two sick calves *in extremis* were found and laboratory examination revealed the presence of a salmonella infection, which was probably largely responsible for the losses in previous seasons.

4. Wild animals can harbour disease agents without showing clinical evidence of disease. The classical example of this so-called carrier state is the infection of various antelopes and equidae with *Tryponosoma* spp. in Africa without showing any harmful effects. They do, however, serve as a source from which *Glossina* spp. can transmit the parasite to domestic animals and cause a fatal disease. A further example is the ability of the warthog and bushpig to act as symptomless carriers of the African swine fever virus. The virus of malignant catarrhal fever of bovines is harboured by wildebeest without the latter showing clinical evidence of disease. The zebra may serve as a source of infection for African horsesickness and certain antelopes for bluetongue in sheep. Wild animals probably serve as reservoirs for the Rift Valley fever virus of man and domestic ruminants. In this way the diseases mentioned may be maintained as inapparent infections in wild animals in the complete absence of man and his domestic stock.

5. Wild animals display a much higher resistance to some diseases than domestic animals, e.g. anaplasmosis, babesiosis and theileriosis. Apart from the fact that such diseases occur in a very mild form, the phenomenon offers vast unexploited possibilities of selecting disease resistant breeds for domestication.

When a parasite contacts an animal population, a host-parasite relationship results. This condition may be delicately balanced and forced in either direction by external influence, e.g. factors causing stress in the host. In order to study the vast variety of host-parasite relationships in existence among wild life, the following methods, classified according to the type of parasite concerned, are recommended. It should be mentioned that the list of parasites is by no means comprehensive as there are still extensive gaps in our knowledge as stated earlier. Parasites in the various classes to which wild herbivora are known to be susceptible will be dealt with as being representative of the class as a whole in respect of the recommended methods of study.

### General considerations

For the investigation of infectious and parasitic diseases in animals, it is essential that suitable specimens are submitted to a competent laboratory for diagnostic examination. Such laboratories exist in most countries and the field officer making the observations and submitting specimens should ensure that such a laboratory, which is able and prepared to receive such specimens, is present in the territory concerned. Furthermore, the taking and submission of specimens and the transport of live and dead animals are governed in most countries by regulations; and, again, the responsible person should make himself conversant with the situation in the country concerned. For these reasons close liaison with the veterinary authorities is essential.

Specimens submitted should be accompanied by a detailed description of all relevant facts in connection with the case, including the locality, species of animal(s) concerned, numbers of animals sick and/or dead, age of animals, symptoms shown, findings on post mortem examination and a list of specimens submitted. It is important that the specimens collected should be in accordance with the disease or diseases suspected.

It should further be borne in mind that, if circumstances permit, it is frequently of greater value to submit the entire carcass of a recently dead animal or a sick animal for examination rather than preserved specimens.

**Some of the more common preservatives used**

**Specimens in 10% formalin for histological examination**

(1) The only fixative to be used is 10% formalin. This is made up by diluting 1 part of commercial formalin (37—40% formaldehyde) with 9 parts of normal saline or tapwater.

(2) The material must be fresh. Take the specimens as soon as possible after the death of the animal.

(3) *The specimens must not be larger than $\frac{1}{2}$ in $\times$ $\frac{1}{2}$ in $\times$ $\frac{1}{2}$ in in order to ensure complete fixation.*

(4) Cut the tissue blocks with a sharp knife or pair of scissors so as not to bruise the tissue. Do not handle more than is necessary and place immediately into the 10% formalin.

(5) The 10% formalin solution in the bottle must completely cover the tissue blocks. There should be at least 10 volumes of the formalin solution to every one volume of tissue. Use two or three bottles rather than overfill one bottle.

(6) Do not use very old solutions of 10% formalin, especially in the case of rabies specimens, as some of the formalin may have evaporated and the tissue blocks will not be properly fixed.

(7) When specimens are submitted from cases which had shown nervous symptoms clinically, or from cases where lesions are suspected to be present in the nervous system (with the exception of rabies), the entire brain and spinal cord should be forwarded in 10% formalin, whenever possible.

In such cases the brain and spinal cord must be fixed in a large volume of 10% formalin; one gallon for bovine and equine brains, and half a gallon for the brains of smaller animals. The brains of the larger species should be allowed to harden in the formalin for 24 hours. They should then be cut transversely into three equal parts with a sharp knife and returned to the formalin to fix. If the brain is cut into three parts while still fresh and soft before hardening in formalin, the examination of it is made more difficult; if a large brain is not incised or if a small quantity of fixative is used, autolysis and putrefaction will occur which, of course, render the specimen useless for examination. It is not necessary to cut the brains of the smaller species (sheep etc.) into three parts; they may be fixed *in toto*.

The brain specimen should either be allowed to fix for 48 hours before being mailed in order to prevent unnecessary damage to it during transit while still relatively soft, or it should be allowed to fix for 10 days and then forwarded in a greatly reduced volume of 10% formalin.

The anterior quarter of the cerebral hemispheres may be removed for bacteriological and virological specimens when necessary.

Specimens for the diagnosis of rabies should be dealt with as indicated in the relevant section.

(8) Eyes should be fixed *in toto* with the optic nerve intact.

**Specimens in 50% glycerine solution for the examination of some diseases caused by bacteria and viruses**

(1) The preservative is a sterile solution of equal parts of glycerine and normal saline. This solution may have a bactericidal or viricidal action on some bacteria and viruses; and it is, therefore, not an ideal general purpose preservative.

(2) Specimens must, if possible, be taken immediately after the death of the animal or, preferably, immediately after a sick animal has been slaughtered. Invading putrefactive organisms often make a diagnosis impossible.

(3) *Specimens should be taken under sterile conditions with sterile instruments.*

(4) The specimens as a general rule should be 1 in × 1 in × ½ in in size.

(5) They must be completely covered by the 50% glycerine solution. There should be at least 2 parts of the solution to every 1 part of the cut sections.

(6) Portions of the stomach, intestine or intestinal contents must *not* be placed in the same bottle with the specimens from other organs. Use a separate bottle for this purpose.

**3% Boracic acid/normal saline solution.** This preservative is used only for blood or serum samples for testing for the presence of antibodies of certain diseases. It consists of a saturated solution of boracic acid (approximately a 3% solution) in normal saline. About 1 ml of this solution should be added to each 9 ml of blood collected.

**O.C.G. solution.** This consists of a sterile mixture of 8240 ml water, 5000 ml glycerine, 50 ml carbolic acid and 50 gm potassium oxalate.

Use 1 part O.C.G. to every 2 parts of blood collected. Collect the blood in as sterile a manner as possible.

**Specimens packed in ice.** Specimens should be collected in as sterile a manner as possible and placed in sterile, wide-mouth, screw-capped bottles which should be as small as possible, e.g. a McCartney bottle. Specimens of organs should be about 2 in × 2 in × 1 in in size and should, if possible, be taken immediately after the animal has died or, preferably, after it has been slaughtered. (*See also* Specimens in 50% glycerine above.)

A wide-mouthed thermos flask is the best to use. Wrap the specimen bottle in cotton wool, tie the wool firmly in place with string (this is to prevent the specimen bottle from breaking the glass of the thermos flask after the ice has melted), place in the thermos flask and then fill the flask to the top with ice.

Keep the specimens in a refrigerator until it is convenient to despatch them in the thermos flask. If specimen bottles containing specimens from more than one case are included in one thermos flask, they must be marked or numbered in a waterproof fashion. Ordinary gum labels become detached in water.

A tin of camping refrigerant packed with a specimen in an insulated container is effective for keeping a specimen chilled for one day or longer.

### Specimens for toxicological examination

Specimens must be submitted in accordance with the actual poison suspected and should be accompanied by the reasons for such suspicions. *Specimens must be submitted in thoroughly cleaned bottles* (*preferably fruit jars and NOT tins*). *Do not fill the bottles more than three-quarters full.*

A label must be attached to each bottle stating:
(a) name and address
(b) contents
(c) preservative (if any)
(d) test desired

### (1) Mineral Poisons

ARSENIC

No preservative is necessary but 10 ml (2 teaspoonsful) of chloroform per jar retards decomposition.

(*a*) *Acute poisoning*
  For *large animals* submit the following:
(i) approximately $1\frac{1}{2}$ lbs ruminal contents and $1\frac{1}{2}$ lbs abomasal contents in separate containers.
(ii) $1\frac{1}{2}$ lbs of liver in a third container.
  For *small animals:*
(i) The whole stomach with contents. If the animal is very small, send also the intestines with the stomach in the bottle.
(ii) Approximately $1\frac{1}{2}$ lbs of liver or the whole liver in a separate container.
  For *birds:*
(i) The contents of the crop, gizzard and proventriculus in one bottle.
(ii) The liver in another bottle.

(*b*) *Chronic poisoning*
  Submit those specimens listed under (a) above plus the following: 2 lbs of skin and 2 lbs of bone (preferably the femur), all in separate containers.

(*c*) *Suspected arsenical poisoning in live animals*
  Submit:
    (i) 4 oz. of hair (cut from close to body)
    (ii) about 1 pint of urine
    (iii) 1 lb of dung.
  All must be placed in separate containers. It is advisable to collect and submit urine and faeces collected on 3 or 4 different occasions in the course of 1 or 2 days.

(*d*) *Materials suspected of causing the poisoning,* feeds, soil scrapings etc. Submit 1 lb or more.

(*e*) *Exhumation*
  Specimens of bone, skin and, if available, the stomach wall and/or stomach contents and liver are required as well as soil samples taken from near the carcass. If possible, submit at least 2 lbs of each.

COPPER POISONING
  No preservative is necessary, but specimens can be preserved in formalin (excepting stomach contents) or 10 ml of chloroform may be added to each jar.

LEAD POISONING

No preservative is necessary but specimens can be preserved in 10% formalin (excepting stomach contents) or 10 ml of chloroform may be added to each jar.

Submit specimens as for arsenical poisoning listed above, plus the kidneys in a separate container. Also include, if possible, specimens taken from the suspected source of poisoning (1 lb or more).

NITRATES AND NITRITES

Submit only very fresh specimens of ruminal contents diluted with an equal volume of alcohol. Submit also the material which is suspected of having caused the poisoning (about 2 lbs).

SODIUM CHLORIDE

No preservative is required but 10 ml chloroform per jar retards decomposition. Submit 1½ lbs stomach contents, 1½ lbs liver and the kidneys in separate containers. If possible, also send 2 lbs of the suspected source of poisoning.

## (2) The nerve poisons

STRYCHNINE

Strychnine is resistant to decomposition itself but its presence may be masked by the products of decomposition giving false negative results on analysis.

Submit the following:

(i) The entire stomach and contents. The stomach must be tied off at both ends thus enclosing the contents. Immerse completely in 96% ethyl alcohol, i.e. rectified spirits and *NOT* methylated spirits.

(ii) The liver must also be included. Cut into strips ¼ in. in thickness and completely immerse in *ethyl alcohol* in a separate container.

## (3) General

(a) *Prussic acid poisoning*

The following specimens are required:

(i) Ruminal contents, liver and muscle—200 g of each in separate containers. Use one pint fruit jars, containing 200 ml of a 1% mercuric chloride solution, as containers. Add specimens until the volume has doubled, i.e. the specimens are just well covered with the preserving fluid. The liver and muscle should be cut into strips about ¼ in. in thickness immediately before being placed in the preservative. Take the specimens immediately after death of the animal, as dydrocyanic acid is volatile and escapes easily.

(ii) Submit suspected plants for identification.

(iii) *The quantitative determination of the cyanide content of fresh plant material.* It is essential that the exact weight or volume of the 1% mercuric chloride (corrosive sublimate) solution and that of the specimen be noted. Use equal amounts of each (about 200 g). Hay and leguminous pods may be submitted dry with no preservative if this is the state in which they are normally fed.

*(b) Dicoumarol*

Submit only very fresh specimens of liver, kidneys and stomach contents in separate containers and well covered with *ethyl alcohol* (and *NOT* methylated spirits).

*(c) Synthetic insecticides*

Poisoning with chlorinated hydrocarbons such as D.D.T., B.H.C., Dieldrin, Toxaphene and organic phosphates, such as parathion and systox, can only be confirmed by complex tests on fresh material. Submit specimens of the liver and stomach contents in separate containers. The specimens should be kept well chilled.

**Plant or fungal poisonings** (Mycotoxicoses)

Where a plant or fungal poisoning is suspected do not submit animal tissues unless the suspected plant or fungi produces specific lesions determinable by histological examination. Such specimens should then be submitted in 10% formalin, plus a pressed, mounted specimen of the suspected plant or material contaminated by fungi for identification (*see* below).

For example:

(i) Specimens of the liver in 10% formalin for seneciosis, crotalariasis, lupinosis, poisonous algae, aflatoxicosis.

**Poisonous plants and fungi for identification**

*Plants must be sent with roots and bulbs, leaves, and, if possible, flowers and fruit.* Especially where the Liliaceae family is concerned, submission of the bulb and leaves only is useless, as there are no identifiable characteristics when flowers are absent. In the case of bigger plants the flowers with a twig of leaves should be submitted.

After collection the plant or part thereof must immediately be spread out on a sheet of blotting paper or newspaper and covered with another sheet. Place between two pieces of fairly rigid cardboard and secure as tightly as possible with string.

If more than one plant is submitted, it is advisable to number them separately. Similarly numbered duplicates should be kept by the sender for reference purposes. If possible, mention its local name and give a short description of it, e.g. whether a shrub, a tree, creeper, etc. perennial, height, the type of soil and veld where it occurs, etc.

## Diseases caused by viruses

### General considerations

#### (1) INFECTIVE MATERIAL

In many virus infections the virus has disappeared from the blood by the time clinical symptoms other than those due to fever appear. *Therefore specimens from which a virus may be isolated must, as a rule, be collected as early as possible in the course of the disease.*

Where possible, *infective material* should be packed in ice and forwarded as rapidly as possible. However, if this is not possible, 50% glycerine solution will have to suffice except in the case of blood specimens.

#### (2) ACUTE AND CONVALESCENT PHASE SERA
#### FOR ANTIBODY DETECTION

*When collecting serum for the detection of specific antibodies, remember that they usually develop slowly. Wherever possible, acute and convalescent phase serum must be submitted.* Serum should be collected as early in the course of the disease as possible and again after an interval of about four weeks *from the same animal* if it recovers. In this way a rise in the antibody level can be detected. Single samples of serum taken after recovery give no indication of length of time that antibodies have been present.

When collecting blood for the detection of specific antibodies, one has two alternatives:

(i) Bleed into a sterile bottle. Allow the blood to clot. Decant the serum into another sterile bottle. Submit packed in ice.

(ii) As for (i) above but after decanting the serum add 500 international units of penicillin and 500 micrograms of streptomycin per millilitre of serum. It is not essential, but is preferable, to pack in ice. This method is preferable to (i) above. A full history of the disease is essential.

## Diseases caused by bacteria and fungi

### General

Specimens for bacteriological examination should only be sent if they can be obtained fresh, collected by the necessary aseptic technique, and preserved and transported so as to assure that they will arrive in good condition.

There is no object in sending semi-putrid or contaminated material, as no significance can be attached to the result. Such specimens will be destroyed.

If examination for more than one pathogen is required, then the requisite number and type of specimens should be sent (e.g. Joint Fluid for *Corynebacterium* spp. and P.P.L.O. examination or intestine for *Cl. welchii* spp. and *E. coli*).

*Note: Never send intestinal specimens together with other organs in the same bottle.*

### Preservation of specimens for bacteriological examination

(i) *Ice*

Specimens to be examined for the presence of fastidious organisms should, where possible, be sent in a thermos flask preserved with ice or dry ice. They should be placed in sterile, 1 oz wide-mouthed McCartney bottles, which should be marked properly and wrapped in cotton wool in order to prevent breaking of the thermos flask should the ice melt. Plastic bags should not be used.

Exudates, pus, milk and small organs should also be sent on ice.

(ii) *50% glycerine solution*

Glycerine kills the organisms in the superficial parts of the specimen. Specimens sent in 50% glycerine solution should therefore be relatively large —about 2·5 × 2·5 × 1·5 cm in size in order to prevent complete penetration. Specimens smaller than the size suggested should be sent on ice.

A 50% glycerine solution is suitable for the less fastidious type of organisms, such as *Enterobacteria* and *Clostridia*.

(iii) *Enrichment media*

Specimens sent in Selanite F or Tetrathionate broth should not exceed 10% of the volume of the medium.

**Sera for serodiagnosis**

Serum should be collected by bleeding into a 10 ml centrifuge tube, allowing the blood to clot at 37° C for 2 hours and then centrifuging at 3000 xg for 15 minutes.

Recentrifuge the supernate, if necessary, and decant the clear serum into a sterile rubber stoppered bottle and dispatch immediately. A few grains of merthiolate may be added to the serum but this is generally unnecessary.

## Collection of entomological specimens

(1) *Ticks*

Send in a solution of 70% alcohol plus 5% glycerine in 5% formalin.

Live ticks for investigational work should be sent in a glass vial or test tube plugged with cotton wool. If live ticks have to travel for more than 24 hours, care must be taken not to send too many per vial.

(2) *Mosquitoes and similar insects*

These can be sent between thin layers of tissue paper in a container or, if facilities exist, pinned in specimen boxes.

(3) *Fleas and lice*

These should be sent in suitable containers with tissue paper to prevent movement or in 70% alcohol (methylated spirits 3 parts, water 1 part constitutes a suitable preservative).

(4) *Flies*

These may be sent dry (as for mosquitoes) and preferably pinned.

(5) *Skin scrapings*

Skin scrapings for the diagnosis of scab or mange should be sent dry in a suitable container e.g. test tube or envelope and, if possible, biopsy material in 10% formalin for histological examination.

## Collection of specimens of helminths

(1) *Examination of faeces for ova*

About 1 oz ($\pm$ 1 tablespoonful) of fresh faeces (collected from the animal) is required. The faeces may be submitted without a preservative providing it reaches the laboratory within 24 hours. Should it take longer, $\frac{1}{2}$ teaspoonful phenothiazine must be mixed into the faeces.

(2) *Nematodes*

If alive when collected, place them in 20 to 30 ml water and add a few drops of Lugol's iodine to kill them. Add 2—3 ml formaldehyde solution.

If already dead, place in 10% formalin.

(3) *Trematodes*

If alive when collected, place in a bottle containing water. Screw on lid. Shake well. Remove lid; add 10% formalin by volume. Replace lid immediately. Shake well again.

If already dead, place in 10% formalin.

(4) *Cestodes*

If alive when collected, drop them into hot alcohol-glycerine *or* into water heated to 70° C, then replace water with cold alcohol-glycerine. Alcohol-glycerine consists of 95 ml 75% alcohol and 5 ml glycerine. If alcohol-glycerine is not available, methylated spirits may be used as follows: 3 parts methylated spirits and 1 part water plus 5% glycerine.

## Technique for post-mortem examination

The technique to be described is that applied to the post-mortem examination of the equine. This is basically the same for all mammals, with modifications according to anatomical variations in the different species of herbivores.

## Examination of body and body cavities

In general, all external or superficial observations must be recorded before distributing or cutting up the carcass.

*Characteristics:* Species, age, sex.

*External description:* Condition.

*Skin:* Appendages, swellings in and under skin, external parasites.

*Natural openings* and their mucous membranes.

*Manipulation:* Remove skin wherever possible, with animal lying on its right side. Remove udder in the female by a complete circular incision. In the male, remove the external genital organs as far back as the ischial arch. Remove the left fore-limb, muscles of the back and lateral aspect of the thoracic cavity on the left side.

*Description: Blood:* Staining quality, fluidity, etc. *musculature. Subcutaneous connective tissue and fat depots. Superficial lymph glands.*

*Manipulation:* Open the abdominal cavity by making a longitudinal incision from the xiphoid cartilage to the pubis followed by a transverse incision from the umbilicus to the angle formed by the last rib and vertebrae.

*Description: Abdominal cavity:* Content (fluid and parasites) and position and displacement of viscera.

*Manipulations:* Open the thoracic cavity by incisions through:
(1) the muscular portion of the diaphragm on the left side from the lumbar region to the xiphoid region,
(2) the costal cartilages on the left side, commencing at the caudal aspect,
(3) the dorsal extremities of the ribs on the left side commencing at the last rib.

*Description: Pleural cavity:* parietal pleura (i.e. costal, diaphragmatic, mediastinal, pericardial), thymus, and bronchial and mediastinal lymph glands.

## Exenteration of organs

Exenterate organs in the following sequence, in general taking the uppermost first:
1  *Spleen*
2  *Small intestine*
3  *Small colon*
4  *Left kidney*
5  *Large colon*
6  *Stomach*

7  *Liver, right kidney, etc.:* Remove liver together with pancreas, duodenum, periportal glands, right kidney and right adrenal in one operation starting from the cranial aspect.

8  *Pericardial sac:* Open the pericardial sac by an incision in the direction of its long axis. Carefully lift the heart by the apex, without wasting any fluid that may be present. Describe contents and parietal pericardium.

9  *Heart:* Describe content, parietal pericardium, state of ventricles (e.g. disatole, rigor mortis, etc.). Open right ventricle by an incision parallel to the descending branch of the right coronary artery; open the right atrium by a transverse incision to admit the hand. Describe content of right ventricle and atrium. Open the left ventricle by an incision parallel to the descending branch of the right coronary artery. Open the left atrium by a transverse incision to admit the hand.

Hold the heart at the apex and remove it from the thoracic cavity by cutting through the vessels at its base.

10  *Lungs:* Examine the pleural surfaces and remove both lungs together with the thoracic and cervical parts of the trachea and oesophagus by severing from their attachments.

11  *Larynx, pharynx and tongue:* Sever their attachments to the medial surfaces of the mandibles and the skull and bucchal cavity.

12  *Brain:* Remove the skin and as much as possible of the subcutaneous tissues of the head. Using a saw, expose the brain by making the following incisions:

(a) transverse incision across frontal region passing through the oral aspect of the temporal fossa

(b) longitudinal incision on both sides of the skull extending from the foramen magnum to the orbital fossa. The plane of this incision should pass just medial to the occipital condyle and the supra-orbital process of the temporal bone.

By means of a chopper or chisel the cranial cap thus formed can be prised up and removed. Incise the dura mater over the hemispheres and dissect out the brain.

13  *Hypophysis*

14  *Nasal cavities:* Make a median longitudinal incision through the skull. This is best made with the saw, finishing with the chopper through the teeth. Disarticulate the mandible and expose the teeth.

15  *Spinal cord:* Remove the flesh from the dorsal aspect of the vertebral column. Then with the chisel, chopper or bone forceps remove the dorsal

spinous processes of the vertebrae and as much as possible of the dorsal wall of the spinal canal as will allow the cord to be dissected out, without injuring the cord itself.

16 *Bone marrow of long bones:* Remove the humerus, femur and longest rib from the skeleton and cut away the flesh. Fix each in the vice separately and with the saw make a sagittal incision which may be finished with chisel or chopper.

## Examination and description of exenterated organs

### 1. Heart:

*Description:* Size, shape, epicardium.

*Manipulation:* Connect up the incisions made into the right atrium and ventricle. From the apex, cut the lateral wall of the right ventricle about ½ inch from the right descending coronary sulcus and so upwards through the pulmonary artery.

*Description:* Right atrial and ventricular endocardium, fossa ovalis, right atrioventricular valves, semi-lunar valves.

*Manipulation:* Proceed in the same way with the left side of the heart. Make an incision flatwise through the substance of the myocardium.

*Description:* Cut surface, colour, consistency, etc.

### 2. Lungs:

*Description:* Shape, size, inflated or deflated, colour of surface, pleura weight.

*Manipulation:* With the lung lying on its costal surface make a transverse incision about 4 inches from the caudal end. Open and examine the bronchi and pulmonary vessels. Palpate for nodules or other consolidations, making longitudinal slices where necessary.

*Description:* Cut surface, edges, colour, scrapings, and consistency.

### 3. Liver and pancreas:

*Manipulation:* Open the duodenum lengthwise and describe contents and mucous membrane. Remove the duodenum, pancreas, and periportal lymph glands. Open up the bile ducts; note contents.

*Description: Pancreas:* size, shape, colour, and consistency. *Liver:* size, shape, colour of surface capsule (visceral and parietal), cut surface (edges, scraping, colour structure) and consistency.

**4. Spleen and splenic lymph glands:**
*Description: Splenic lymph glands:* size, shape, cut surface, colour and scrapings.
*Manipulation:* Remove splenic lymph glands.
*Description: Spleen:* Size, shape capsule, cut surfaces, edges, pulpa, trabeculae, malpighian bodies, scrapings, consistency.

**5. Kidneys and adrenal glands:**
*Description: Adrenal gland:* Size, shape, colour, cut surface, cortex, medulla.
*Kidneys:* Fat capsule.
*Manipulation:* Remove fat capsule.
*Description:* Size, shape, colour.
*Manipulation:* With the kidney lying flat make a horizontal incision through its substance, commencing at the lateral border of the kidney and carrying the incision to near the hilus.
*Description:* Cut surfaces, edges, scrapings, colour, structure.
*Manipulation:* Remove fibrous capsule.
*Description:* Fibrous capsule, surface of kidney, consistency.

**6. Tongue and cervical organs:**
*Description: Tongue:* Surface and cut surface, palpation of musculature.
*Lymph glands:* retropharyngeal, mandibular, cranial, and cervical: size, shape and cut surface.
*Salivary glands:* ditto.
*Thyroid and parathyroids:* ditto.
*Pharynx and tonsils:* Surrounding tissues, cavity, mucous membrane.
*Manipulation:* Open the oesophagus and trachea by a longitudinal incision along the dorsal wall.
*Description: Oesophagus and Trachea:* (cervical portion). Contents, mucous membrane and vocal chords.

**7. Thoracic aorta and trachea:**
*Manipulation: Thoracic aorta and surrounding tissues:* Open the vessel by a longitudinal incision.
*Description: Thoracic aorta and surrounding tissues:* Elasticity and intima.
*Thoracic trachea:* Contents and mucous membrane.
*Thoracic oesophagus:* Contents, mucous membrane and musculature.

**8. Abdominal aorta, cranial mesenteric, etc.:**

*Manipulation: Abdominal aorta and surrounding tissues:* Open the vessel by a longitudinal incision.

*Cranial mesenteric and its branches and surrounding tissues:* Open by long incision.

*Description: Abdominal aorta and surrounding tissues:* elasticity and intima.

*Cranial mesenteric and its branches and surrounding tissues:* Contents, elasticity and intima.

*Mesenteric lymph glands:* Size, shape, cut surface, colour, scrapings.

*Mesentery:* Fat content and colour.

**9. Stomach:**

*Description:* Serosa, size, and shape.

*Manipulation:* Open the stomach by an incision along the greater curvature.

*Description:* Contents and mucous membrane.

**10. Small intestine and small colon:**

*Description:* Serosa, size and shape.

*Manipulation:* Open along the attachment of the mesentery.

*Description:* Contents and mucous membrane.

**11. Large colon and caecum:**

*Description:* Serosa, size, and shape.

*Manipulation:* Open, commencing with the small colon and carrying the incision through the right dorsal portion, left dorsal portion, left ventral and right ventral portions of the large colon and through the caecum. The incision is to be made about 2—5 cms from the attachment of the mesentery.

*Description:* Contents and mucous membranes.

**12. Bladder, sexual organs, rectum, etc.:**

*Description:* Size and shape; description of each organ after opening as to contents, mucous membranes, etc.

**13. Brain and hypophysis:**

*Description:* External surface: blood vessels, depth of sulci, character of gyri and pia mater, origin of the cranial nerves, pons, and medulla.

*Manipulation:* Place the brain on its basal surface with the medulla towards operator. Open the left ventricle by making an incision just above the corpus callosum. Expose and examine the contents of the left ventricle hippocampus, fimbria, corpus striatum and plexus of the left lateral ventricle. Turn the brain round so that the oral aspect of the cerebrum now faces the operator. Proceed in the same way with the opening and examination of the right ventricle, etc. By careful dissection expose the thalami, epiphysus and corpora quadrigemina. Open the third ventricle by a longitudinal incision through the thalami. Note the contents and the opening of the ductus aqueductus. Open the fourth ventricle by an incision through the middle of the vermis of the cerebellum. Note and examine the contents and the opening of the central canal of the spinal cord. Make a medial longitudinal incision through the stem of the brain and examine the cut surface. Make sagittal incisions through the hemispheres of the cerebrum and lobes of the cerebellum and examine the cut surfaces.

*Description: Hypophysis:* Size, shape, colour and cut surface.

**14. Cranial cavity, dura mater, spinal cord, cranial and spinal nerves:**
*Description:* Tumors and condition.

**15. Nasal cavities, sinuses and guttural pouches:**
*Description:* Contents, mucous membranes, contents and parasites.

**16. Teeth:**
*Description:* Degree of wear and condition.

**17. Bone marrow of humerus and femur:**
*Description:* Color and consistency of the marrow.

**Ruminants and other species**

In ruminants a similar technique is to be followed with the following modifications.

(1) *Removal of the forestomachs together with the spleen*
   After removal of the left forelimb the abdominal cavity is opened by a longitudinal incision extending from the xiphoid cartilage to the pubis.

Turn the animal over onto the left side and make a transverse incision on the right side through the umbilical and iliac regions extending as far as the transverse process of the lumbar vertebrae. Ligate and sever the duodenum in two places: (a) near the pylorus, (b) under the right kidney. Detach as much as possible of the omentum on the right side, the right ruminal artery, the 2 branches of the omasoabomasal arteries, and as much of the attachment of the dorsal aspect of the rumen on the left side as possible. Turn the animal back onto its right side and make a transverse incision through the umbilical and ilial regions, extending as far as the transverse processes of the lumbar vertebrae on the left flank.

*Remove the omentum* altogether and *proceed with the removal of the forestomach and the spleen by:*
(a) Ligating and severing the oesophagus.
(b) Cutting through the adhesions of the rumen on the left side.
Remove the left lateral wall of the thorax and then the rest of the abdominal viscera with the following modifications:

(2) *Removal of the small intestine*
Ligate and sever the ileum about 1 inch from the *ileocaecal* opening and proceed cranially with the removal of the small intestine by cutting through the attachment of the mesentery close to the bowel wall. The liver is removed together with attached loop of duodenum intact.

(3) *Removal of colon and caecum*
Ligate and sever the large intestine at its entrance to the pelvic cavity. Proceed cranially with the removal of the large intestine intact. Cut through the mesentery and dissect away the whole of the large intestine without disturbing the coils of the ansa spiralis. Now proceed with the examination and removal of the thoracic organs, etc., as in the equine.

(4) *In the examination of the liver, pancreas, etc., proceed in ruminants as follows:*
(a) Open duodenum and note whether ductus choledochus is patent.
(b) *Pancreas:* Size, shape, cut surface.
(c) *Periportal lymph glands:* ditto.
(d) *Gall bladder:* Size, shape, contents, mucous membrane.
*Manipulation:* Remove the duodenum, pancreas, periportal lymph gland and gall bladder and proceed with the examination of the liver as in the equine.

(5) *Removal of the brain:* Make the following incisions:
  (a) Transverse incision through the aboral aspect of the orbital fossa.
  (b) Longitudinal incision along the median plane of the frontal region, extending from the foramen magnum to the transverse incision orally. By applying pressure to and using the horns as levers, the cranial cavity can be exposed. In hornless animals proceed as for the equine. In other animals the same general principles apply with minor modifications according to anatomical differences.

## Post mortem report or protocol

All the findings of properly conducted post-mortem examinations should be recorded in writing. In the body of the report are given the details of the appearance of the carcass and organs examined. These should be in *plain descriptive language* and not in the form of *diagnoses*. Under *Pathological Anatomical Diagnosis* a summary of the findings is given so as to reflect all the gross abnormalities found. This should be very concise, preferably couched in brief technical diagnostic phrases or terms. These should be given or arranged in the following order:

(a) Extent of post-mortem changes, if present.
(b) State of carcass in general, e.g. the conditions of the animal, cachexia, anaemia, presence of icterus; infiltrations of connective tissues.
(c) Various systems, lymphatic and blood vascular system (including the heart), respiratory, urogenital, digestive and nervous systems, etc.

Attention should be paid to the correct terminology of the pathological alterations, e.g. acute or chronic, catarrhal, haemorrhagic, fibrinous, etc., nature of the inflammatory processes, petechiae, ecchymoses, peritoneal or subpleural haemorrhagic extravasations, etc. It is desirable in all cases that an opinion of the nature of the changes be formed. If, however, doubt exists, full descriptions should be given and further investigation, e.g. chemical examination, microscopical, carried out to clinch the diagnosis.

Finally, an opinion must be expressed regarding the actual cause of death i.e. *etiological diagnosis*. In connection with this, as much qualifying information as possible should be given.

## Collection of specimens

As stated above, suitable material for further investigation must be collected
during the course of post-mortem and properly preserved and labelled at
once. Such material would include:

(1) Pathological specimens: For histological section, organ smears, museum
    pieces, etc.
(2) Specimens for chemical analysis: Stomach contents, liver, etc.
(3) Internal and external parasites for identification.
(4) Any other material which may be of interest or of use in formulating or
    supporting the diagnosis, foreign bodies, samples of suspicious food or
    water supply, poisonous plants, etc.

# 15

# Conversion Factors for Productivity Measurements

GEORGE A. PETRIDES

The conversion factors beyond have proved useful in studies of the energy relations of organisms. Many items listed are available elsewhere but are included here for convenience.

These were among data presented in September, 1965, at the IBP Symposium on Large Herbivores in Aberdeen, Scotland.

TABLE 15.2. Areas of plots permitting easy conversion of productivity data (with radius for circular plots of such size).

Plot sizes:

| | | | | |
|---|---|---|---|---|
| ·001 acre | = | 43·560 | square feet | (44·70 inches) |
| ·0001 acre | = | 4·356 | square feet | (14·14 inches) |
| ·001 hectare | = | 10·000 | square meters | (1·784 meters) |
| ·0001 hectare | = | 1·000 | square meter | (0·564 meters) |

Kilograms per hectare are equal to the number of:

| | | | |
|---|---|---|---|
| grams per | 0·924 | square feet | (7·529 inches) |
| grams per | 10·000 | square meters | (1·784 meters) |
| ounces per | 3·797 | square feet | (1·099 feet) |
| ounces per | 1377·865 | square meters | (20·94 meters) |

Pounds per acre are equal to the number of:

| | | | |
|---|---|---|---|
| grams per | 96·033 | square feet | (5·529 feet) |
| grams per | 88·922 | square meters | (1·685 meters) |
| ounces per | 2722·5 | square feet | (29·44 feet) |
| ounces per | 252·937 | square meters | (8·973 meters) |

TABLE 15.1. Multiplication factors for converting productivity measurements.

*Areas:*

| | | | | | | | | |
|---|---|---|---|---|---|---|---|---|
| Acres | × | 43,560 | = | Square feet | × | $2296 \times 10^{8}$ | = | Acres |
| Acres | × | 0·00156 | = | Square miles | × | 640 | = | Acres |
| Acres | × | 0·4047 | = | Hectares | × | 2·4710 | = | Acres |
| Square feet | × | 0·0929 | = | Square meters | × | 10·471 | = | Square feet |
| Square yards | × | 0·836 | = | Square meters | × | 1·1962 | = | Square yards |
| Square miles | × | 259·004 | = | Hectares | × | 0·00386 | = | Square miles |
| Square meters | × | 10,000 | = | Hectares | × | 0·0001 | = | Square meters |

*Weights:*

| | | | | | | | | |
|---|---|---|---|---|---|---|---|---|
| Ounces | × | 28·35 | = | Grams | × | 0·03527 | = | Ounces |
| Pounds | × | 435·592 | = | Grams | × | 0·0022 | = | Pounds |
| Pounds | × | 0·4526 | = | Kilograms | × | 2·2046 | = | Pounds |
| Pounds | × | 0·0005 | = | Short (US) tons | × | 2000 | = | Pounds |
| Pounds | × | $4·4642 \times 10^{-4}$ | = | Long (British) tons | × | 2240 | = | Pounds |
| Pounds | × | $4·5359 \times 10^{-4}$ | = | Metric tons | × | 2204·6 | = | Pounds |
| Grams | × | 1000 | = | Kilogram | × | 0·001 | = | Grams |
| Kilograms | × | 1000 | = | Metric ton | × | 0·001 | = | Kilograms |
| Ounces per square foot | × | 305·167 | = | Grams per square meter | × | 0·00328 | = | Ounces per square foot |
| Pounds per acre | × | 1·1208 | = | Kilograms/hectare | × | 0·8922 | = | Pounds per acre |
| Pounds per acre | × | 0·0001121 | = | Kilograms/square meter | × | 892·2 | = | Pounds per acre |
| Pounds per square mile | × | 0·00175 | = | Kilograms/hectare | × | 571·4 | = | Pounds per square mile |
| Pounds per square mile | × | $175 \times 10^{-9}$ | = | Kilograms/square meter | × | 5,714,285·71 | = | Pounds per square mile |
| Kilograms per hectare | × | 0·1 | = | Grams/square meter | × | 10 | = | Kilograms per hectare |

# Index